TOOL, MACHINE, & EQUIPMENT

SAFETY AND OPERATION

Thomas A. Hoerner
Professor Emeritus
Dept. of Agriculture & Biosystems Engineering and Agriculture Education & Studies
Iowa State University Ames, Iowa

Mervin D. Bettis
Professor Emeritus Department of Agriculture
Northwest Missouri State University
Maryville, Missouri

Revised by T.J. Brown
South Central College
North Mankato, Minnesota

ROWMAN & LITTLEFIELD
Lanham • Boulder • New York • London

Please email textbooks@rowman.com and request access to instructor resources.

Credits and acknowledgments for material borrowed from other sources, and reproduced with permission, appear on the appropriate pages within the text.

Published by Rowman & Littlefield
An imprint of The Rowman & Littlefield Publishing Group, Inc.
4501 Forbes Boulevard, Suite 200, Lanham, Maryland 20706
www.rowman.com

86-90 Paul Street, London EC2A 4NE

First Rowman & Littlefield paperback edition 2023

First Printing 1973
Second Printing 1975
Third Printing 1977
Fourth Printing 1980
Fifth Printing 1984
First Revision 1987
Seventh Printing 1990
Eighth Printing 1994
Second Revision 1998
Tenth Printing 2001
Eleventh Printing 2005
Twelfth Printing 2008

British Library Cataloguing in Publication Information available

Library of Congress Cataloging-in-Publication Data
Library of Congress Control Number: 2023935955

∞™ The paper used in this publication meets the minimum requirements of American National Standard for Information Sciences—Permanence of Paper for Printed Library Materials, ANSI/NISO Z39.48-1992.

TOOL, MACHINE, & EQUIPMENT SAFETY AND OPERATION

INTRODUCTION

This workbook, *TOOL, MACHINE, & EQUIPMENT SAFETY AND OPERATION,* has been prepared for instructors and students in agricultural education, technical education, trades & industry and other agencies for safety instruction in the areas of tool, machine, and equipment safety and operation. *TOOL, MACHINE, & EQUIPMENT SAFETY AND OPERATION* is a much revised and expanded version of *POWER TOOL SAFETY AND OPERATIONS: WOODWORKING, METALWORKING, METALS AND WELDING* published by Hobar Publications in 1998. In addition, this workbook would be valuable to a homeowner aiding in the safe and proper operation of power tools and equipment commonly found in the home shop.

The workbook is designed to give the student a basic understanding of the proper operation of tools, machines and equipment. A variety of tools, machines, and equipment are covered that are commonly found in school laboratories/shops. For each tool, machine, or piece of equipment, there is a section on part identification, safe operational procedures, general safety practices, and completion questions. After the student has read the written material and completed the part identification and completion question section, the instructor should determine the proper adjustment and operational features of the power tool being studied. Following this, the student should successfully complete the safety exam and then be allowed to operate the power tool with close supervision from the instructor. The next step in this educational process is for the student to complete an activity of their choosing using the power tool or tools that have been studied. Note: As manufacturers' designs differ for the same power tool, make certain the operator's manual is available and used for making adjustments and operation of the specific tool, machine or piece of equipment.

An Instructor's Packet designed to be used in conjunction with this manual is available. It contains suggested instructor and student activities; tool, machine, and equipment diagram masters; and the safety exams. This packet is very useful in the teaching of this material.

Almost everyone involved in school laboratory/shop instructional programs will come in contact at some time with tools, machines and equipment. The safe and proper operation of these tools, machines, and equipment is essential for quality and efficient student work. Knowledge on the safe and proper operation of tools, machines and equipment will make the work easier and more enjoyable and will increase operator satisfaction and performance.

Dedication

The writing team, and specifically the project lead, is especially grateful to Rowman & Littlefield for the opportunity to revise, modernize, and expand one of the best educational safety resources of the last fifty years. Updating the original 30 power tools to include modern cordless versions of the same equipment and then adding 50 more power tools and machines was a tremendous endeavor. Each individual sacrificed a great deal of time away from their families and local schools and programs to complete this labor of love, all in an effort to help young people learn to use power tools, machines, and equipment in a safe manner. We received a great deal of support from the following individuals at the Minnesota Department of Education: Joel Larsen, Tim Barrett, and Zane Sheehan; and the following individuals at South Central College, North Mankato, MN: Gale Bigbee, Brad Schloesser, Shelly Kitzberger, Brianna Salfer, Wes Taylor, and Taylor Kong.

Revision Writing Team

Alison Almos-Anoka Hennepin Schools

Luke Becker-Braham High School

Bob Bonin-Fairmont High School

T.J. Brown-South Central College

Tim Hahn-Elk River Schools

Ashley Johnson-Hinckley Finlayson High School

Dean Joslin-Anoka Hennepin Schools

Mark Lockhart-New Prague High School

Hannah Reisdorf-Moorhead High School

Timm Smith-Foley High School

Tom Spehn-NE Metro 916

Revision Project Lead

Thomas "T.J." Brown started his teaching career at Springfield High School (MN) delivering both agricultural education and industrial technology courses for twelve years. Being a small school and a single teacher department, the courses offered in the laboratory/shop included general middle school, woods, residential construction, metals, mechatronics, small engines, auto mechanics, and CAD, so a knowledge of teaching with a broad spectrum of tools, machines, and equipment was useful for this project. Currently, at South Central College, North Mankato, MN, his coursework includes agribusiness, welding, facility maintenance, workplace environment layout & planning, and mechatronics. In addition to teaching, T.J. and his wife Mandy are parents to three lovely daughters, actively involved in their church, and enjoy taking family trips to the lake.

TABLE OF CONTENTS

AIR COMPRESSOR

PART IDENTIFICATION

Identify the circled parts on the air compressor illustrated below.

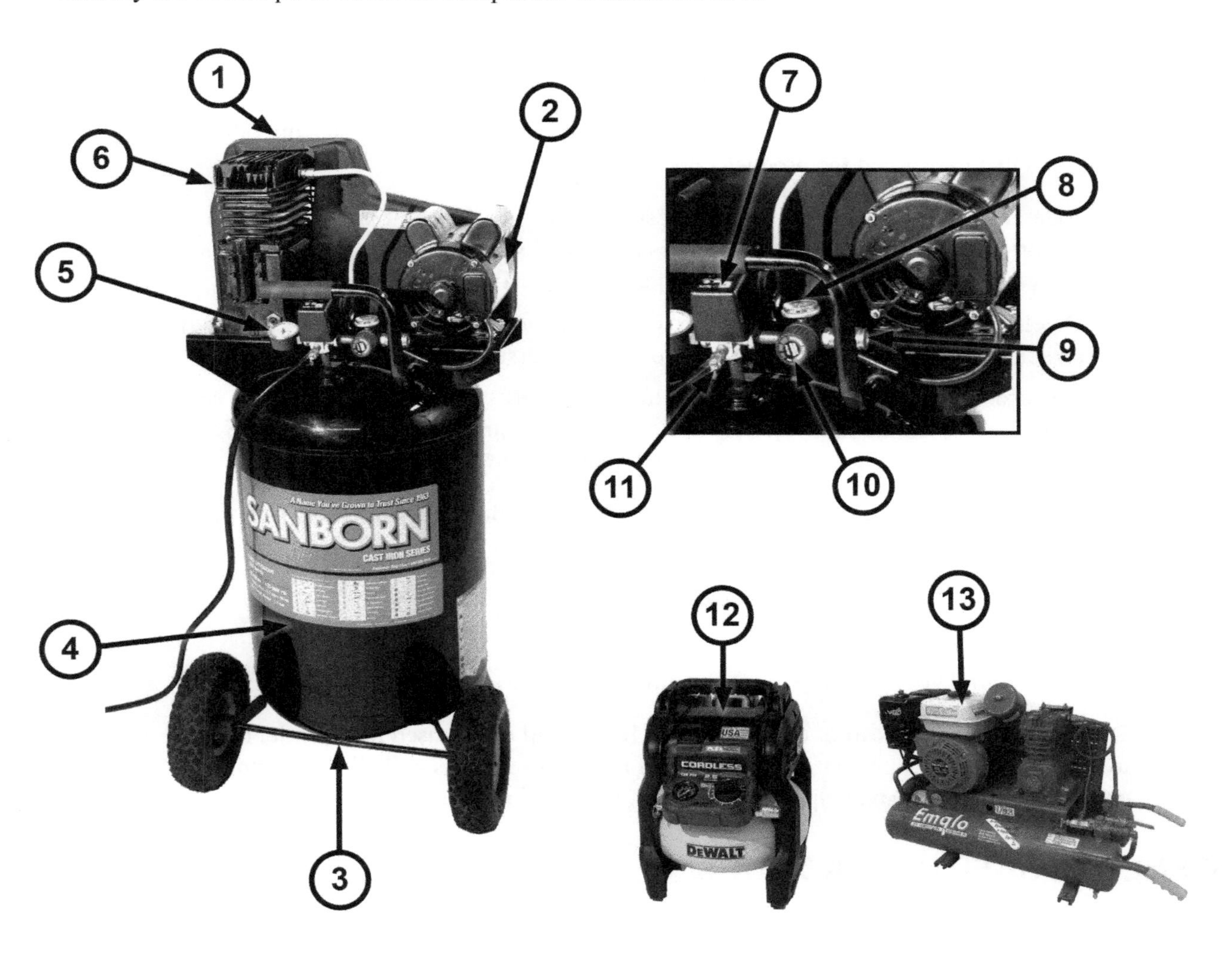

1. ______________________________

2. ______________________________

3. ______________________________

4. ______________________________

5. ______________________________

6. ______________________________

7. ______________________________

8. ______________________________

9. ______________________________

10. ______________________________

11. ______________________________

12. ______________________________

13. ______________________________

SAFE OPERATIONAL PROCEDURES

1. Always operate the compressor on the floor in a level position. All splash-lubricated pumps have a dipper on the bottom of the connecting rod that must remain submerged in the oil bath.
2. Check all oil levels daily and change at regular intervals. Refer to the maintenance section of the owner's manual for the correct type/weight of oil to use and how often the oil should be changed.
3. Check that the drain valve is closed at the bottom of the tank before starting.
4. Check for misalignment or binding of moving parts, breakage of parts, and any other condition that may affect the product's operation.
5. Energize machine and allow the air compressor to run. Watch the air gauge and make sure the compressor switches off when it reaches the model's rated pressure capacity. This will generally be around 100-120 pounds per square inch (PSI).
   a. On electric air compressors, plug cord end into appropriate wall outlet. Flip the power switch to the ON position and allow the air compressor to run.
   b. On gasoline powered air compressors, turn the on/off switch to the on position and if equipped, turn the fuel valve to the on position. Adjust the choke lever to the recommended setting and pull the recoil handle to start the engine.
6. Use the air control valve or regulator to adjust the PSI of the air compressor. This should be based on the tool you'll be using.
7. Attach the air hose to your air compressor, make sure you have enough hose to safely reach your work area.
8. Connect the tool you will be using to the other end of the air hose.
9. Drain tanks of moisture after each day's use. If unit will not be used for a while, it is best to leave drain valve open until such time as it is to be used. This will allow moisture to completely drain out and help prevent corrosion on the inside of tanks.

GENERAL SAFETY PRACTICES

1. Wear approved eye protection, hearing protection (if required), and proper clothing. Tie up loose hair and remove loose jewelry.
2. Do not operate the air compressor without the instructor's permission, or without instructor supervision.
3. Never add or change the oil or refuel when the compressor is running or has just recently been used.
4. Check the safety "pop off" or safety relief valve by manually pulling on the attached ring. The pop off valve will discharge compressed air in the event the regulator is not operating correctly. Release the ring to return to normal operation.

5. Do not operate air compressor in hazardous atmospheres, such as in the presence of flammable liquids, gases, or dust. Power tools create sparks which may ignite the dust or fumes.
6. Operate air compressor in an open area at least 18 in. away from any wall or object that could restrict the flow of fresh air to ventilation openings. Do not use a gas-powered air compressor indoors.
7. Don't expose air compressors to rain or wet conditions. Water entering an air compressor will increase the risk of electric shock.
8. Do not abuse the cord. Never use the cord to carry the air compressor or pull the plug from an outlet. Keep cord away from heat, oil, sharp edges, or moving parts. Replace damaged cords immediately. Damaged cords increase the risk of electric shock.
9. When operating an electric air compressor outside, use an outdoor extension cord. These cords are rated for outdoor use and reduce the risk of electric shock.
10. Do not exceed the pressure rating of any air hose or tool in the system. Protect air hose from damage or puncture. Keep air hose and power cord away from sharp objects, chemical spills, oil, solvents, and wet floors.
11. Release all pressures within the system slowly. Flying dust and debris may be harmful.
12. Keep the exterior of the air compressor dry, clean, and free from oil and grease.
13. Inspect tanks yearly for rust, pin holes, or other imperfections that could cause it to become unsafe.
14. Use the air compressor only for its intended use. Do not alter or modify the unit from the original design or function.
15. Always disconnect the air hose and power supply before making adjustments, servicing a product, or when a product is not in use.
16. Always follow all safety rules recommended by the manufacturer of your air compressor. Following this rule will reduce the risk of serious personal injury.
17. Do not attempt to pull or carry the air compressor by the air hose.
18. Use only 3-wire extension cords that have 3-prong grounding plugs and 3-pole receptacles that accept the product's plug.
19. ALWAYS use the lowest pressure that will do the job.

COMPLETION QUESTIONS

1. Check all oil levels _______________ and change at regular intervals.
2. Check that the drain valve is closed at the bottom of the ______________ before starting.
3. Watch the air gauge and make sure the compressor switches ______________ when it reaches the model's rated pressure capacity.

4. Use the air control valve or ______________ to adjust the PSI of the air compressor.
5. Make sure you have enough hose to ______________ reach your work area.
6. Wear approved ______________ ______________, hearing protection (if required), and proper clothing.
7. Do not use a gas-powered air compressor ______________.
8. Release all pressures within the system ______________.
9. Always disconnect the air hose and power supply ______________ making adjustments.
10. Do not attempt to pull or carry the air compressor by the ______________.

SMALL ENGINE

PART IDENTIFICATION

Identify the circled parts on the small engine illustrated below.

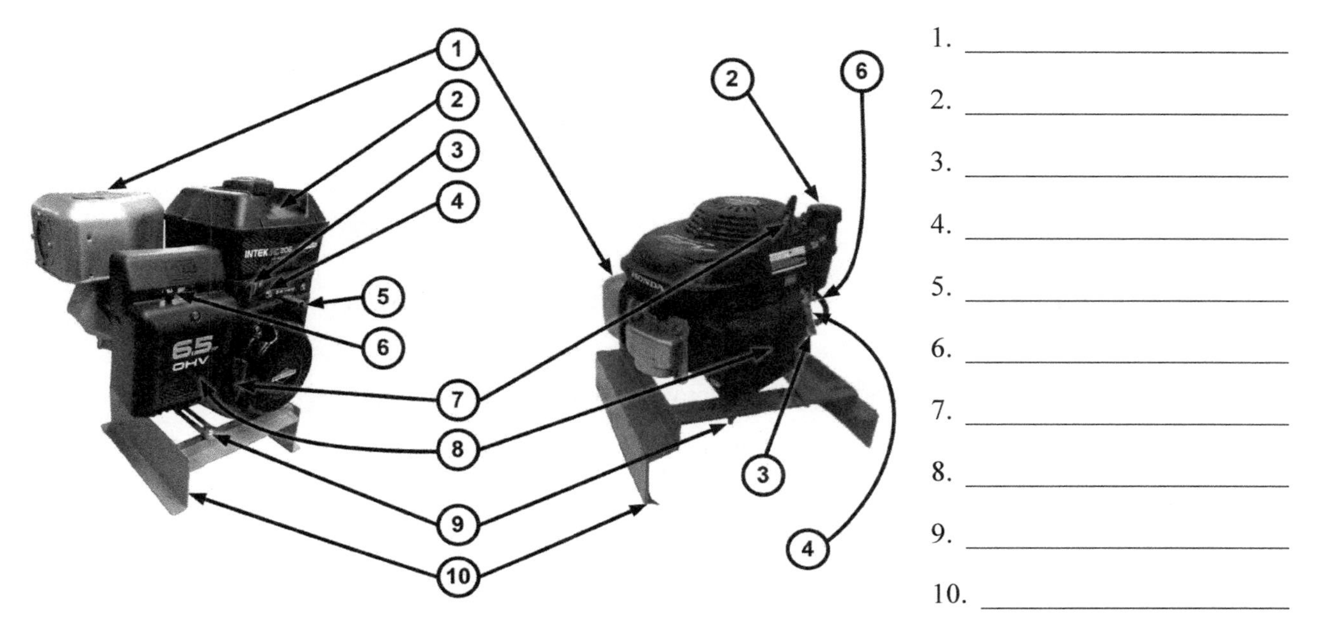

1. ____________________
2. ____________________
3. ____________________
4. ____________________
5. ____________________
6. ____________________
7. ____________________
8. ____________________
9. ____________________
10. ____________________

SAFE OPERATIONAL PROCEDURES

1. Always refer to the owner's manual/operation manual for engine disassembly and engine rebuilding.
2. Gasoline and lubricants are special hazards. The vapors are always present and open flames should never be present in the small engine working area.
3. Always know the location of the fire extinguisher and fire blankets in the work area.
4. Hand and power tools must be used safely to avoid personal injury. Understand proper tool usage and work in a cautious manner.
5. The shop work area must have proper exhaust ventilation to remove fumes from running engines while in the work area.
6. A small gas engine must be in a rigid table mount when being tested, otherwise it can become very unstable while it's running.
7. No engine should be started unless the approved muffler (flame arrester) is in place and in safe working order.
8. Engines get hot when being operated and remain hot for some time after being stopped, it is easy to burn yourself, use caution.

9. No engine should be started unless it has a working throttle cable, an operational shut off (kill) switch or an operational governor.
10. When testing any running engine, have the most current and accurate information about the specific testing procedure you are about to do.

GENERAL SAFETY PRACTICES

1. Wear approved eye protection, hearing protection, and proper clothing. Tie up loose hair and remove loose jewelry.
2. Do not operate the machine without the instructor's permission, or without instructor supervision.
3. In some cases, safety goggles and face shields may be required.
4. Maintain a constant awareness of the many hazards involved with a running small engine on a test stand.
5. Never allow unqualified persons to enter the work area.
6. Ensure the engine clamping devices (bolts) are properly fastened.
7. Be sure that all testing equipment (tachometer, volt meter, spark tester, compression gauge, leak down tester) are safely out of the way when the small engine is running.
8. Do not over rev the small engine, if the RPM gets too high (over 5,000 rpm) the engine may explode.
9. Only one person may operate the testing engine.
10. Keep the work area as clean as possible, be sure to clean up and put away all tools after the engine testing is completed.

COMPLETION QUESTIONS

1. Never operate the ______________ when the instructor is not in the shop.
2. Every ______________ is responsible for helping in the shop clean up.
3. Never use ______________ to clean parts.
4. Ask your instructor for ______________ before starting an engine.
5. Check fuel lines for ______________.
6. Vent engine exhaust to the ______________ ______________.
7. Keep your hands away from ______________ ______________.
8. Do not run an engine ______________ than its rated RPM.
9. When running an engine at high speed, wear face and ______________ protection.
10. Drain ______________ from an engine before it is stored in the shop.

AIR CHISEL/IMPACT HAMMER

PART IDENTIFICATION

Identify the circled parts on the air chisel/impact hammer illustrated below.

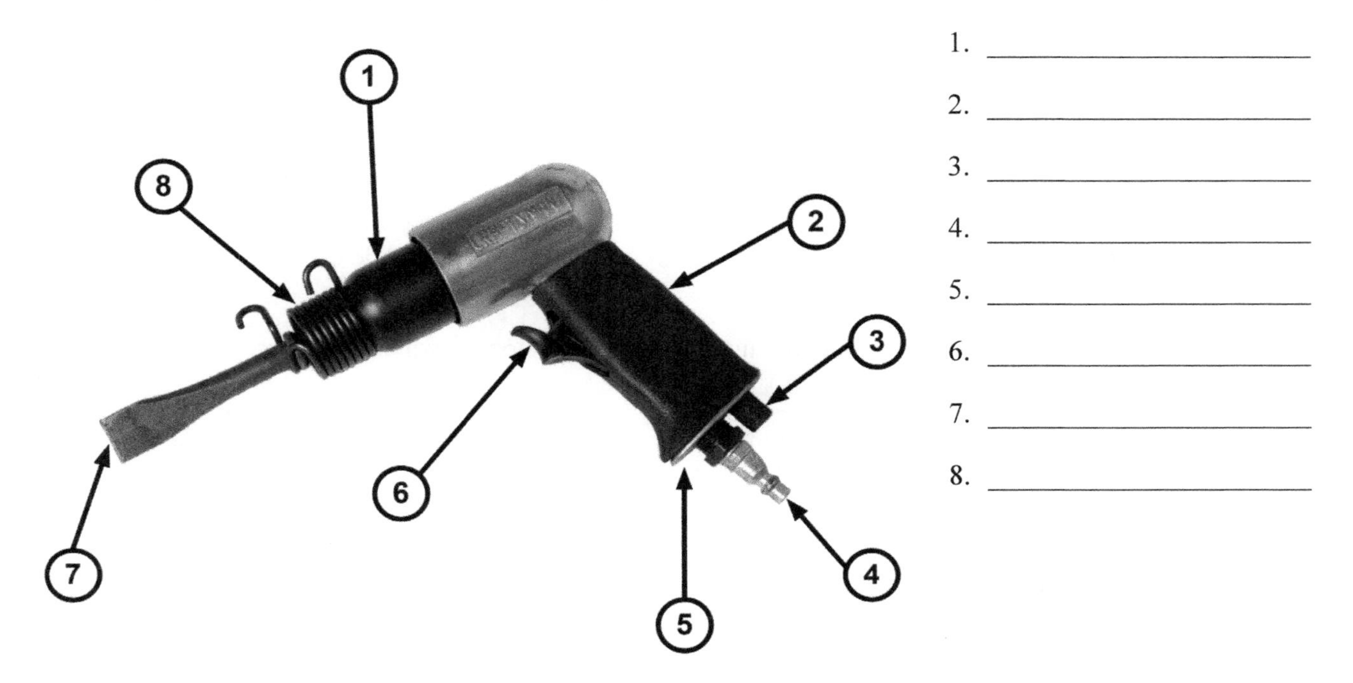

SAFE OPERATIONAL PROCEDURES

1. Review the manufacturer's information and instructions before using the air chisel/impact hammer tool.
2. When using the air chisel, fatigue and strain may occur from continuous use. This can cause the user to have the tools slip or fall from their hands.
3. Flying debris is a risk when using the air chisel, be sure to wear safety glasses and or a face mask.
4. Firm footing is required of any operator of an air chisel/hammer.
5. Be sure all your work is secured with clamps or tightly in a bench vise.
6. Short bursts of air-power should be used to operate all air tools.
7. Using any air tool requires the use of quality hearing protection, ear muffs are strongly advised. Others in the work area should also have ear protection.
8. "Raymond Syndrome" is when your hands and fingers go numb or lose feeling. This can be caused by the vibrations created from using any air tool too long.
9. Always wear protective work gloves while using the air chisel.

10. Put all waste and metal trimmings in the proper receptacles.
11. Store the air chisel/hammer tool in a secure, safe and clean area.
12. Keep the air chisel well lubricated (because of water in the air hose), every time you use it.
13. The air tool should be disconnected from the air hose at the tool.
14. Make sure the shop's compressed air supply is clean and dry.
15. Warn others in your work area that you will be using the air chisel, they can then get their personal protective equipment.

GENERAL SAFETY PRACTICES:

1. Wear approved eye protection, hearing protection, and proper clothing. Tie up loose hair and remove loose jewelry.
2. Do not operate the machine without the instructor's permission, or without instructor supervision.
3. To follow the safety precautions recommended by the manufacturer, and do not exceed the manufacture psi rating for the tool.
4. To have proper ventilation when using an air chisel
5. Do not use compressed air to blow debris from clothes, 5 psi could rupture an ear drum, or debris could be blown into the eyes.
6. If the air tool is not working properly, you should tell your instructor.
7. Always hold firmly onto the air chisel/impact hammer with both hands, applying steady and even pressure, but never hold on to the cutting bit.
8. Check hoses regularly for cuts, bulges, and abrasions; tag and replace if defective.
9. Do not operate the air chisel/impact hammer above the manufacture pressure rating.
10. Hardened chisels and punches are designed for use in air tools, ensure you have the proper bit for the job.
11. Do not carry the air tool by the hose.
12. Avoid tripping hazards caused by hoses laid across walkways.
13. Use the proper hose and fittings of the correct diameter.
14. Choose air supply hoses designed for high pressure pneumatic tools, which are rated at a higher psi than what the tool and compressor are rated.
15. Make sure air chisel/impact hammers are stored in a clean, dry and safe area, wipe the bits with a clean rag before storing, never use soap and water.
16. Alert others in the work area that you are operating a tool that will require ear protection.
17. Turn off the air pressure to hose when not in use or when changing power tools.

18. Bits should be discarded when they become dull, never resharpen.

COMPLETION QUESTIONS

1. Be sure all your work is _______________ with clamps or tightly in a bench vise.
2. _______________ footing is required of any operator of an air chisel/hammer.
3. _______________ chisels and punches are designed for and should be used in every air tool.
4. The air tool should be disconnected from the air hose at the _______________.
5. _______________ bursts of air-power should be used to operate all air tools.
6. If the air tool is not working properly, you should _______________ _______________ _______________.
7. Always hold firmly onto the air chisel/hammer with _______________ _______________.
8. You should apply straight, steady and _______________ _______________ when using an air tool.
9. Use an air hose specifically designed for high pressure _______________ tools.
10. Do not use _______________ air to blow debris or clean dirt from clothes.

AIR NOZZLE

PART IDENTIFICATION

Identify the circled parts on the air nozzle illustrated below.

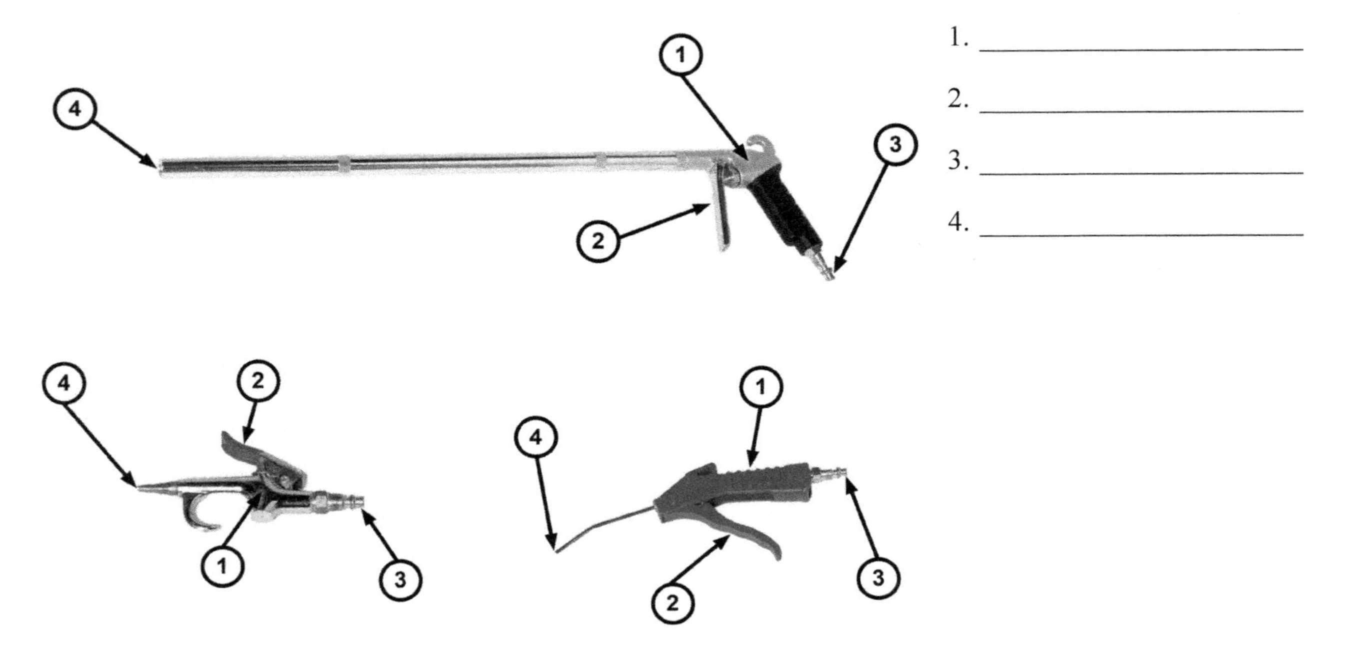

SAFE OPERATIONAL PROCEDURES

1. Make certain that you have approved PPE required for the task.
2. Some air nozzles have an assortment of tips that can be attached to the end of the air nozzle. Verify that you have secured the correct tip for the task prior to installing the air hose.
3. Inspect air hose for any visible signs of wear prior to connecting to air nozzle. Do not use an air hose that has signs of wear and report to an instructor.
4. Verify that you have the same fitting style and size that connects the air nozzle to the air hose.
5. Depending on fitting type, either push the fitting (attached to the air nozzle) into the coupler (attached to the air hose), or you may have to pull back the sleeve on the coupler before inserting the fitting into the coupler.
6. Once connected, make sure you have no air leaks at the coupler and the connection is secure. Report any air leaks to the instructor.
7. If used for cleaning, you cannot exceed 30 psi.
8. Place the tip of the air nozzle approximately 18 inches away from the object and squeeze the trigger to allow compressed air to flow through tip.

9. After compressed air is flowing through tip, move air nozzle as close to the object as necessary to obtain required results.
10. To stop the flow of compressed air, release the trigger.
11. When finished with the air nozzle remove it from the hose in the opposite way you attached it.

GENERAL SAFETY PRACTICES

1. Wear approved eye protection, hearing protection, and proper clothing. Tie up loose hair and remove loose jewelry.
2. Do not operate the machine without the instructor's permission, or without instructor supervision.
3. Wear a face or dust mask if the operation is dusty.
4. Keep your work area clean and well lit. Cluttered benches and dark areas invite accidents. Floor must not be slippery from wax or dust.
5. Do not operate air nozzle in explosive atmospheres, such as in the presence of flammable liquids, gases or open flames.
6. Keep bystanders, children, and visitors away while operating an air nozzle. Distractions can cause you to lose control.
7. Do not overreach. Keep proper footing and balance at all times. Proper footing and balance enable better control of the air nozzle in unexpected situations.
8. Protect air hoses from damage or puncture. Keep air hose away from sharp objects, chemical spills, oil, solvents, and wet floors.
9. Check hoses for weak or worn condition before each use, making certain all connections are secure. Do not use if defect is found.
10. Always use a clean cloth when cleaning the air nozzle. Never use brake fluids, gasoline, petroleum-based products, or any strong solvents to clean the unit. Following this rule will reduce the risk of deterioration of the enclosure plastic.
11. Make sure the hose is free of obstructions or snags. Entangled or snarled hoses can cause loss of balance or footing and may become damaged.
12. Use the air nozzle only for its intended use. Do not alter or modify the unit from the original design or function.
13. Never leave the air nozzle unattended with the air hose attached.
14. Never point any air nozzle toward yourself or others. The air nozzle should never be used to clean dust from your body or clothing. If compressed air gets into your body through your mouth, nose, ear or skin, it can cause serious injury, such as a ruptured esophagus or eardrum, sudden, permanent hearing loss or even a pulmonary embolism.
15. Do not continue to use air nozzle that leaks air or does not function properly.

16. ALWAYS use the lowest pressure that will do the job.
17. Compressed air shall not be used for cleaning purposes except when reduced to less than 30 psi and then only with effective chip guarding and personal protective equipment. This is to protect the operator and others from the hazards of the release of compressed air and flying debris.
18. Never have air nozzle tip in contact with any object when operating.

COMPLETION QUESTIONS

1. Verify that you have ________________ the correct tip for the task prior to installing the air hose.
2. Do not operate the air nozzle without the ________________ permission, or without instructor supervision.
3. Verify that you have the same fitting ________________ and size.
4. Keep your work area clean and well ________________.
5. Once connected, make sure you have no ________________ leaks.
6. Keep hose away from sharp objects, chemical spills, oil, ________________ and wet floors.
7. Place the tip of the air nozzle approximately ________________ inches away from the object.
8. Check hoses for weak or worn condition ________________ each use.
9. To stop the flow of compressed air, ________________ the trigger.
10. The air nozzle should ________________ be used to clean dust from your body or clothing

AIR/POWER RATCHET

PART IDENTIFICATION

Identify the circled parts on the air/power ratchet illustrated below.

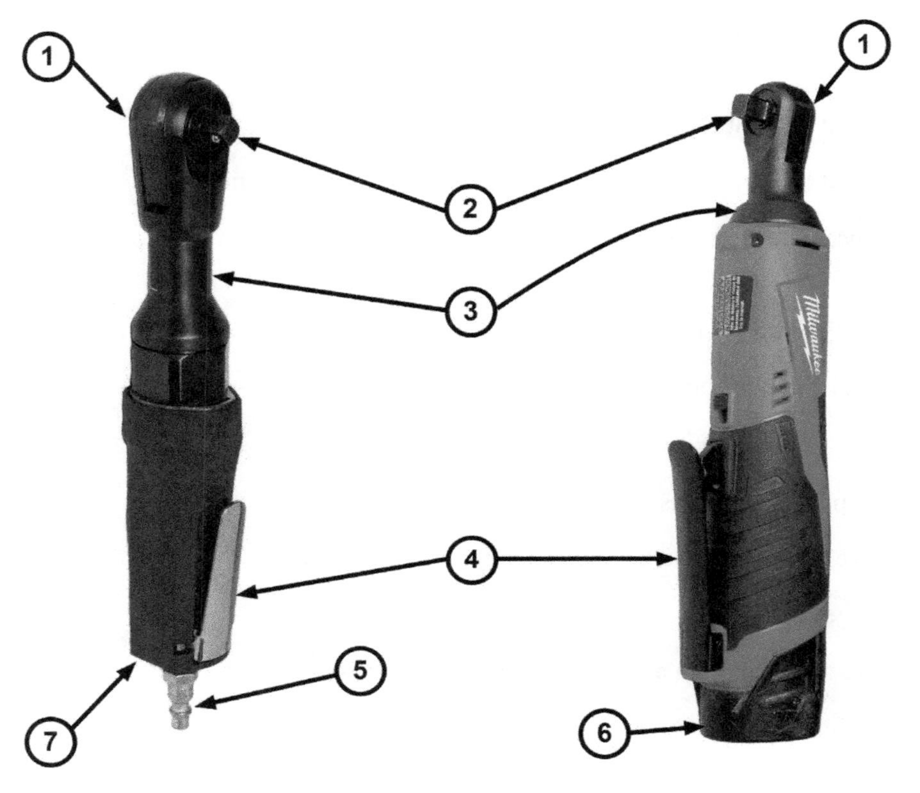

1. ____________________
2. ____________________
3. ____________________
4. ____________________
5. ____________________
6. ____________________
7. ____________________

SAFE OPERATIONAL PROCEDURES

1. Never touch any moving parts of the ratchet while it is spinning.
2. Work in a clean, dry, well-lit and safe environment.
3. Use the right tool for the right job, don't force the tool to do work it's not intended to. An air/power rachet does not have the same power as an impact wrench.
4. Dress appropriately when using the ratchet tool.
5. Keep the ratchet power supply connections clean and dry.
6. If repairs to the tool are needed, contact the tool manufacturer.
7. Use only impact sockets for the air/power ratchet.
8. Use two clean hands and gloves when using the ratchet.
9. Return the ratchet to its proper place, ready for the next person to use.

10. Do not over tighten the nuts and bolts with the power tool, always use a torque wrench after the ratchet, to be sure nuts and bolts are tight to specifications.

GENERAL SAFETY PRACTICES

1. Wear approved eye protection, hearing protection, and proper clothing. Tie up loose hair and remove loose jewelry.
2. Do not operate the machine without the instructor's permission, or without instructor supervision.
3. Do not pass power tools to other students when they are still running.
4. The owner's manual is designed to provide crucial information about how to use the tool.
5. Store in dry, clean and safe space, never use in cold or wet conditions; always clean the tool when not in use.
6. Inspect the tool before every use, read the operator's manual.
7. Identify the size and condition of the nuts and bolts, then get the correct size sockets.
8. Disconnect the power source from the air/power ratchet when the tool is not in use.
9. Notify the instructor if the air/power ratchet is not working.
10. Start the job with caution, have your tools and parts clean and ready for work.
11. Give yourself enough time to get your job done right.
12. Make sure your work is secure and held with a C-clamp or tightly in a vise.
13. Set torque control for correct tightness before starting the job.

COMPLETION QUESTIONS

1. Always use ___________________ type sockets designed for an air/power ratchet.
2. Make sure your work is secure and held with a ___________________ or tightly in a vise.
3. Set ___________________ control for correct tightness before starting the job.
4. Make sure ___________________ hands are free to properly operate the air/power ratchet.
5. Always ___________________ the tool when not in use.
6. The owners ___________________ is designed to provide crucial information about how to use the tool.
7. Having a clean ___________________ is necessary to get the job done correctly.
8. Give yourself enough ___________________ to get your job done right.
9. Always wear ___________________ ___________________, when working on a project.
10. Only use the cordless ___________________ for the jobs it was intended for.

ANGLE GRINDER

PART IDENTIFICATION

Identify the numbered parts of the angle grinder illustrated below.

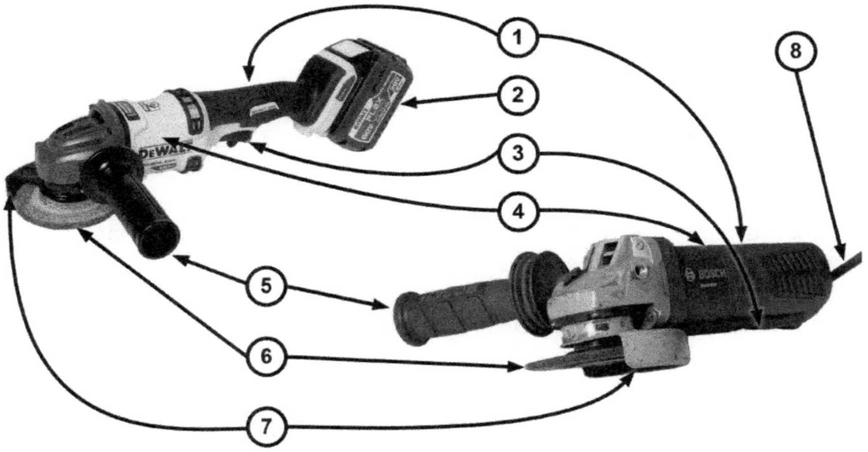

1. ______________________
2. ______________________
3. ______________________
4. ______________________
5. ______________________
6. ______________________
7. ______________________
8. ______________________

SAFE OPERATIONAL PROCEDURES

1. Always check the RPM range of the disc as compared to the speed of the grinder to protect the disc from exploding if too high of an RPM is used.
2. Be sure the nut is tightened correctly on the shaft. If it is upside down, the disc will spin and will not tighten down firmly with the spanner wrench provided. No need to overtighten as the disc will tighten itself during normal operation.
3. Hold the grinder firmly with both hands.
4. Be sure the disc guard is in place. The guard should be between your fingers and the disk to protect your fingers.
5. Be sure no one is in line with the disc before starting the motor.
6. Turn the grinder motor on and off and check to see if the disc vibrates excessively or does not run round and true.
7. When an extension cord is used with the angle grinder, make sure the cord is sufficiently large in size for the grinder being used. A #14 gauge extension cord is minimum and #12 gauge is preferred.
8. Keep the electrical and extension cords positioned away from the grinding area. Cords are easily severed by the angle grinder.
9. Run the grinder at full speed for one minute after a disc has been mounted or has been roughly treated. If a disc is going to break, it will usually do so when the grinder is first turned on.

10. Be sure the work is clamped down firmly or secured in a vise.
11. Feed the grinding disc lightly into the work after the motor has come up to full speed.
12. Do not force the disc into the work and cause the disc speed to be reduced.
13. Do not lay the grinder down until the disc has stopped turning.
14. Lay the grinder on its rest plate so nothing touches the grinding disc while it is not in use.
15. Return the angle grinder to its case or cabinet after use.
16. Identify the location of the tool shut-off for use in an emergency situation.

GENERAL SAFETY PRACTICES

1. Wear approved eye protection, hearing protection, and proper clothing. Tie up loose hair and remove loose jewelry.
2. Do not operate the machine without the instructor's permission, or without instructor supervision.
3. Be sure the switch is off and the cord is disconnected from the power source before making any adjustments, lubricating, inspecting, or changing grinding discs.
4. Never use a cracked grinding disc or one that vibrates excessively.
5. Be sure the grinder is properly grounded or double insulated.
6. Use only discs that are designed to operate at the speed indicated on the grinder nameplate.
7. Be sure the disc guard is in place.
8. Do not direct the discharge at anyone as the sparks can cause burns and small pieces can become embedded in the skin.
9. When grinding small pieces, be sure they are held securely in a vise or clamped to the table.
10. Do not talk to anyone while operating the grinder.
11. Do not use the grinder in areas where flammable materials are kept.
12. Do not wear loose-fitting or frayed clothing.
13. Make sure the floor around the work area is clean and free of other materials before operating the grinder.
14. Keep the power cord away from the grinding disc.
15. Always hold the grinder with both hands.
16. When using the wire brush attachment on the angle grinder, hold the grinder extra firmly to prevent it from being thrown about.
17. Always make sure the switch is off when the grinder is not in use so the grinder will not start when the cord is connected to a power source.

18. Be sure the grinder does not exceed the speed stamped on the side of the grinding disc.

19. Never operate the angle grinder while standing in water or on a wet floor.

20. Never use a grinding disc when it is less than 1/2 of its original diameter.

COMPLETION QUESTIONS

1. The grinder speed should never ________________________ the speed indicated on the side of the grinding disc.

2. The grinder should be lying on its ________________________ when not in use.

3. Do not lay the grinder down until the ________________________ has stopped turning.

4. Hold the grinder with ________________________ ________________________.

5. Be sure the ________________________ guard is in place.

6. Do not ________________________ the disc into the work.

7. Never use the grinding disk when it has been worn past ________________________ of its original diameter.

8. Keep the ________________________ ________________________ away from the grinding wheel.

9. Make sure the switch is in the ________________________ position before plugging it in to the outlet.

10. When grinding small pieces, use a ____________________ or ____________________.

AUTOMOTIVE BATTERY CHARGER

PART IDENTIFICATION

Identify the circled parts on the automotive battery charger illustrated below.

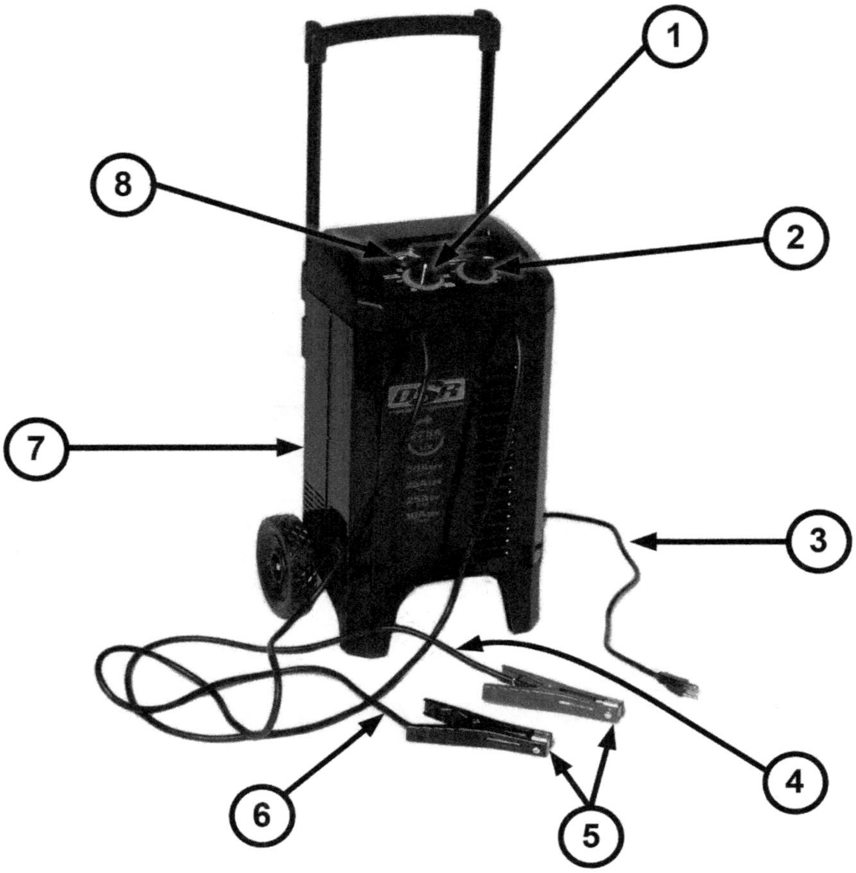

1, ____________________

2. ____________________

3. ____________________

4. ____________________

5. ____________________

6. ____________________

7. ____________________

8. ____________________

SAFE OPERATIONAL PROCEDURES

1. Raise the hood/lid and open the doors of the battery compartment before starting to charge the battery. This will help prevent an explosive mixture of gases building up.
2. Disconnect the vehicle battery cable prior to connecting the battery charger.
3. Before starting to charge a vented (wet cell) battery, check that the electrolyte level is just above the tops of the plates in all the cells. Top off the cells with distilled water if the level is low, and replace the cell covers before charging.
4. Make sure the charger is switched off before connecting the charging leads to the battery (unless the manufacturer specifies a different procedure).

5. Connect the chargers positive (+) lead to the battery's positive terminal and the negative (-) lead to the negative terminal if the battery is not connected to a vehicle.
6. Connect the chargers positive (+) lead to the battery's positive terminal and the negative (-) lead to a ground if the battery is connected to a vehicle.
7. Check to be sure the charging leads are securely clamped in position before switching on the charger.
8. If a spark occurs when the charger is turned on, turn off the charger, check the battery and reconnect the charger clamps.
9. Never charge the battery faster than the battery manufacturer's specified maximum charging rate.
10. Match the manufacturer's recommended amperage charge, do not over charge the battery.
11. Do not remove or adjust the charging leads while the charger is switched on. Always switch it off first.
12. Switch off the charger before disconnecting the charging leads from the battery (unless the manufactures instructions specify otherwise).
13. Allow a vented battery to stand for at least 20 minutes after disconnecting it from the charger. Carefully top off the electrolyte with distilled or deionized water to the recommended level.
14. Store the charging leads so that the uninsulated parts do not rest against each other or any earthed metalwork. This will prevent short circuiting if the charger is switched on suddenly.
15. Acid spills can be neutralized with ammonia, baking soda, or large quantities of water.
16. Always store the automotive battery charger in a clean, dry and safe area.

GENERAL SAFETY PRACTICES

1. Wear approved eye protection, hearing protection, and proper clothing. Tie up loose hair and remove loose jewelry.
2. Do not operate the machine without the instructor's permission, or without instructor supervision.
3. Maintain a constant awareness of the many hazards involved with charging an automotive battery.
4. Never allow the two terminals to come into contact with each other via a conductive material such as a metal tool.
5. If a battery explosion occurs, immediately wash any part of your body with cold water.
6. Never allow unqualified persons to enter the work area.
7. Check the clamping connectors for corrosion, lose or damaged cables.
8. Make sure that all electronic functions of the automobile are turned off before attaching the battery charger.

9. Do not stand directly over a charging battery.
10. Keep the charging battery away from any flames or sparks.
11. Charge the automotive battery in a well vented area.

COMPLETION QUESTIONS

1. Observe general safety rules, always use _______________ _______________ when servicing the automotive battery charging.
2. Keep the charging battery away from any _______________ or _______________.
3. Never allow _______________ persons to enter the work area.
4. Provide ample _______________ when using a battery charger.
5. Disconnect the vehicle battery _______________ cable before charging.
6. Do not stand directly _______________ the charging battery.
7. If a spark occurs when the charger is turned on, _______________ _______________, the charger, check the battery and reconnect the charger clamps.
8. If a battery explosion occurred, immediately wash any part of your body with ____________ ____________.
9. Match the manufacturer's recommended amperage charge, do not ____________ ____________ the battery.
10. Acid spills can be neutralized with ammonia, ____________ soda, or large quantities of water.

AUTOMOTIVE ENGINE LIFT

PART IDENTIFICATION

Identify the circled parts on the automotive engine lift illustrated below.

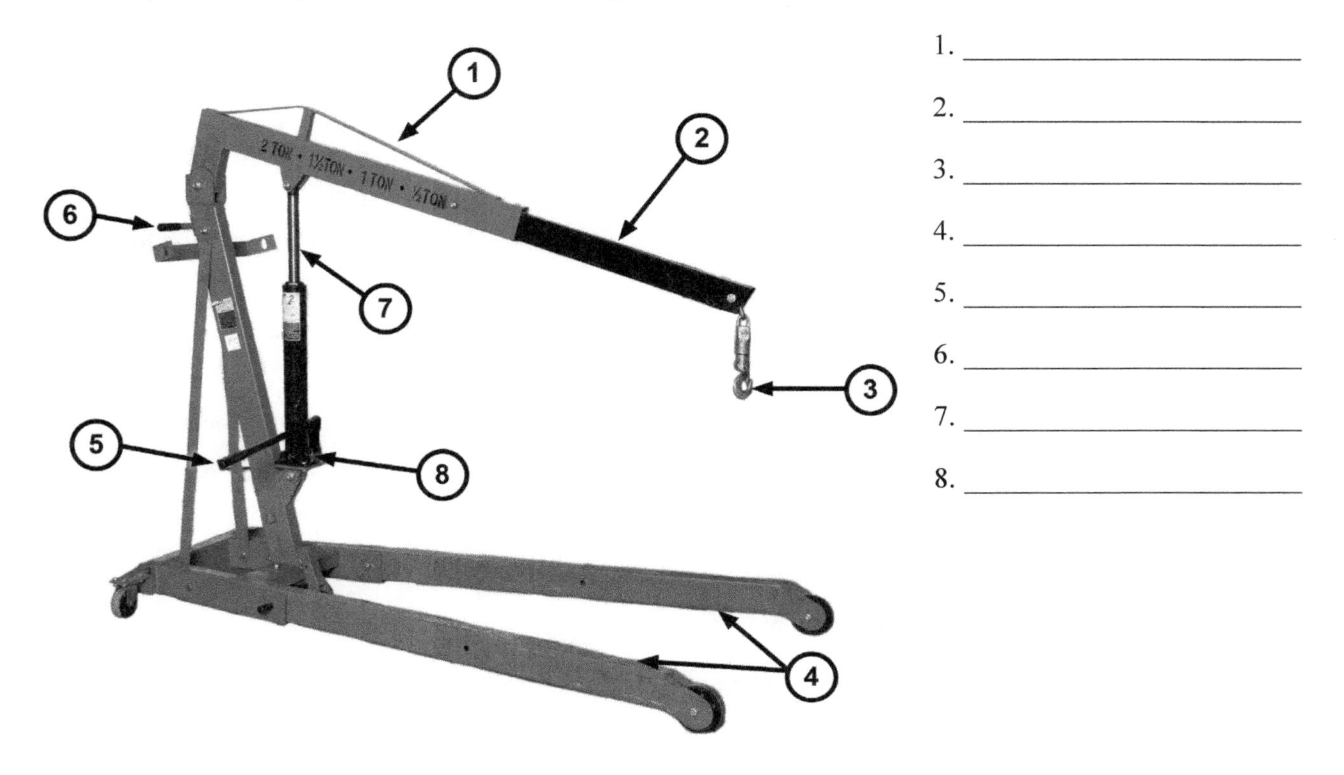

1. ______________________
2. ______________________
3. ______________________
4. ______________________
5. ______________________
6. ______________________
7. ______________________
8. ______________________

SAFE OPERATIONAL PROCEDURES

1. Place the lift directly over the object to be lifted.
2. Inspect the chains, cable, or strap each time it is being used.
3. Attach the lifting device to the object, using no less than 3/8 inch diameter bolts.
4. Double check the fastening of the chains, cable or strap to make sure it is secure and that it cannot slip from the object being lifted.
5. The load must be balanced in the chains, cable, or strap before lifting.
6. Make sure all persons and obstructions are clear before raising or lowering the object.
7. No one is permitted under the vehicle while using the lift.
8. Keep all hands clear and raise the engine until the motor mounts are clear. Shake the engine from side to side to see that it is free.
9. Raise the engine slowly while watching for any binding.

10. After the engine is free from the vehicle, either move the vehicle back or hoist forward until the engine can be lowered to the ground.
11. Do not leave the load in a high position on the hoist.
12. Make sure the size of the load on any lift does not exceed its rated load.
13. It is important to inspect the chain, cable, or strap. Keep them clean and do not use them if they are twisted or damaged.
14. Chains or cables should be oiled and stored in a clean dry place.

GENERAL SAFETY PRACTICES

1. Wear approved eye protection, hearing protection, and proper clothing. Tie up loose hair and remove loose jewelry.
2. Do not operate the machine without the instructor's permission, or without instructor supervision.
3. Maintain a constant awareness of the many hazards involved with lifting engines or heavy objects.
4. Never allow unqualified persons to enter the work area.
5. Check the lifting attachments and chains for corrosion, damage or oily surfaces.
6. Never overload the chains, cable, or strap.
7. Be patient and cautious when lifting an engine or heavy object.
8. Do not use the engine lift for anything but lifting.
9. Never try to stabilize a falling engine, Get out of the way!

COMPLETION QUESTIONS

1. Always read the safety ________________, know how much your engine weighs.
2. Use the engine lift only for ______________ ______________, a job that it is designed for.
3. The lift should only be used by a qualified ________________.
4. Never use the lift for lifting ________________ heavier than allowed.
5. When lifting heavy loads, be sure someone else is ________________ to help if needed.
6. Wear proper ________________ when working with an engine lift.
7. Test the engine lift with a low ________________ load to check to be sure it works correctly.
8. Always check the lift for damaged or worn out ________________.
9. Use the lift on a flat and solid ________________.
10. Do not go under the engine or the ________________ while moving it.

BERNZOMATIC TORCH

PART IDENTIFICATION

Identify the circled parts on the bernzomatic torch illustrated below.

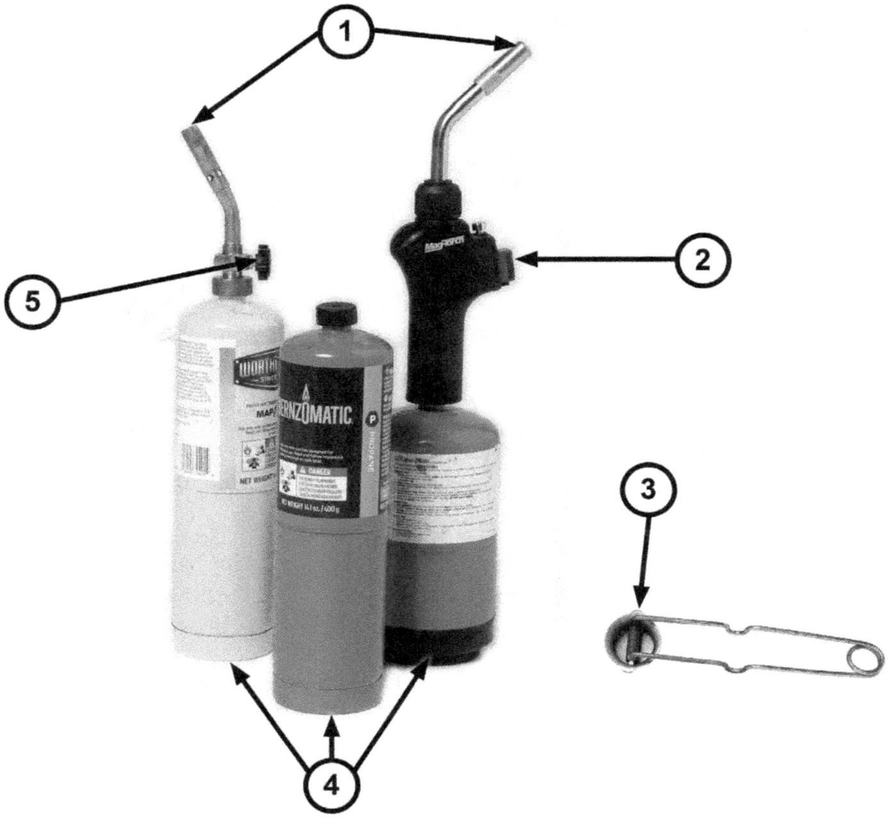

1. ____________________
2. ____________________
3. ____________________
4. ____________________
5. ____________________

SAFE OPERATIONAL PROCEDURES

1. Operation and Lighting (with Instant On/Off)
   a. Before installing the torch to the gas cylinder, turn the ignition button in a clockwise direction to make sure the ignition safety feature is ON.
   b. Before installing the torch to the cylinder, make sure the valve is OFF by turning the gas valve in a clockwise direction until hand tight only. DO NOT FORCE.
   c. Hold torch and gas cylinder upright. Turn the gas cylinder clockwise into the torch valve body until locked tight and attached securely to the torch connector. Check connection for leaks. (Soapy water)
   d. If a leak is detected, replace the cylinder and check for leaks again. If leaking still occurs, the torch must not be used.
   e. Turn the gas valve counter-clockwise about a quarter of a turn. At this stage, the gas is not being released yet.

f. Turn the ignition button counter-clockwise to disengage the ignition safety lock.

g. Squeeze the ignition button, slowly allowing gas flow into the burner tube. Continue to squeeze it until the torch is lit. If torch does not light, squeeze the ignition button again.

h. To engage the ignition button lock for continuous work, pull the ignition button fully inward and press the lock button on top. Then release the ignition button.

2. Operation and Lighting (with a Striker)

   a. Inspect the torch head for signs of damage. If you notice any signs of damage, do not use the torch.

   b. Push the arm of the striker across the striking plate to ensure that the flint located in the striker creates a spark. Refer to the operating instructions for the striker to determine how to install a new flint if you do not get a spark from the striker.

   c. Hold the striker in one hand and turn the gas valve of the torch 1/4 turn to start the gas flowing from the torch head.

   d. Push the striker arm to ignite the torch. It may take a few attempts with the striker to light the torch. Turn off the gas if the torch does not light within five-seconds of you turning on the gas valve and allow the gas to dissipate for 30 to 40-seconds before attempting to light the torch again.

   e. Fully open the gas valve after you have successfully lit the torch before you use it.

3. Flame Adjustment

   a. Turn the valve to control the gas output according to different applications.

   b. The maximum heat zone is concentrated. It is located between ½” and 1” from the end of the tip. Holding the torch too close or too far away from the target can reduce the heating effect.

4. Shutdown and Storage

   a. Release the ignition to extinguish the flame. If the ignition button lock is ON, pull the ignition button fully inward to release the lock.

   b. Turn the valve OFF. Note: The flame may continue to burn for a short period of time if the torch has been operated in an inverted (upside down) position.

   c. When the torch is cool, turn the gas cylinder counter-clockwise to disconnect it. Slightly pull the ignition button to vent the remaining gas inside the torch.

   d. Turn the ignition button clockwise to engage the ignition safety lock.

   e. Store the torch and gas cylinder separately. Read other cautions on gas cylinder label.

5. Cold Weather Operation

   a. In cold weather the size of the flame can be smaller due to lower gas pressure. To produce a larger flame, pre-warm the gas cylinder to room temperature prior to use. Shaking the gas cylinder a few times to warm the fuel up will help. There is a possibility that the flame will extinguish when the torch is turned upside down in cold weather.

GENERAL SAFETY PRACTICES

1. Wear approved eye protection, hearing protection, and proper clothing. Tie up loose hair and remove loose jewelry.
2. Do not operate the machine without the instructor's permission, or without instructor supervision.
3. Wait for the torch to cool down before tightening or removing torch head and before installing or removing accessories.
4. Do not point torch toward face, other people or flammable objects.
5. Disconnect gas cylinder when not in use. Never store with torch attached to the gas cylinder.
6. Read and understand the SDS for all materials before beginning work.
7. Do not attach or detach the torch from the cylinder where accidentally released gas might be ignited by open flames.
8. Never attempt to modify the torch construction and never use unapproved fuel.
9. Do not use a leaking, damaged or malfunctioning torch.
10. Work only in well ventilated areas. Avoid the fumes from fluxes, and all metal heating operations.
11. Always place your work on firebricks.
12. Always wear protective gloves. Use proper tools to handle hot work.
13. Never use the torch on or near combustibles.
14. Always make certain the torch is placed on a level surface when connected to the gas cylinder to reduce the risk of accidental tip over.
15. Always have a fire extinguisher near the torch and work area.
16. Be extra careful when using the torch outdoors on sunny or windy days. Bright sun makes it impossible to see the torch's flame. Wind may carry the torch's heat back towards you or to other areas not intended to be heated. Windy conditions may also cause sparks to be blown into other areas with combustible materials.
17. Never use a torch to strip paint.
18. Heating a surface may cause heat to be conducted to adjoining surfaces.

COMPLETION QUESTIONS

1. After the gas cylinder is attached to the torch, check all connections for ________________.
2. Work only in well ________________ areas. Avoid the fumes from fluxes, and all metal heating operations.
3. Heating a surface may cause heat to be ________________ to adjoining surfaces

4. Always make certain the torch is placed on a _______________ surface when connected to the gas cylinder to reduce the risk of accidental tip over.

5. The maximum heat zone is concentrated. It is located between _______________ and _______________ from the end of the tip.

6. Never attempt to _______________ the torch construction and never use_______________ fuel.

7. Wait for the torch to _______________ _______________ before tightening or removing torch head and before installing or removing accessories.

8. Always place your work on _______________.

9. Store the torch and gas cylinder _______________. Read other cautions on gas cylinder label.

10. Never use the torch on or near_______________.

BISCUIT JOINTER

PART IDENTIFICATION

Identify the circled parts on the biscuit jointer illustrated below.

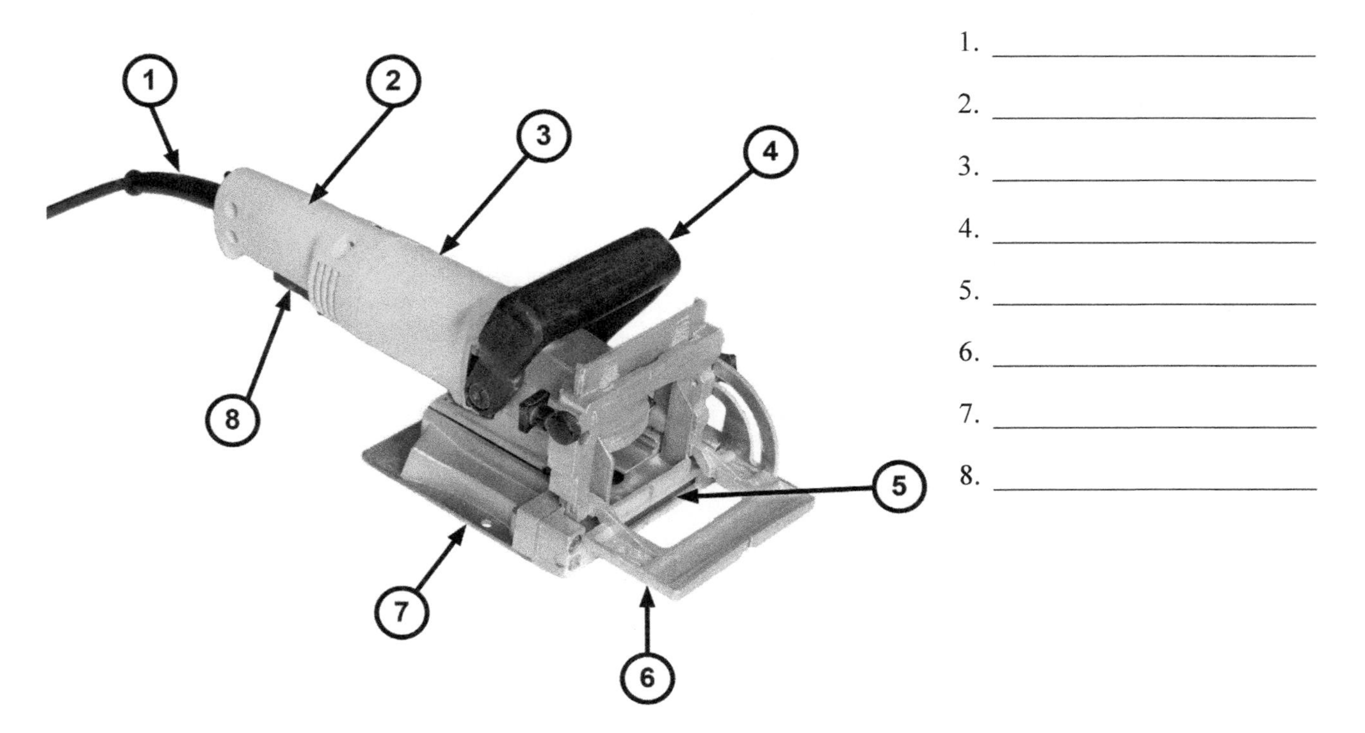

SAFE OPERATIONAL PROCEDURES

1. Inspect the material for loose knots, nails or other hardware, do not cut into these.
2. Do not biscuit joint material less than 3/8".
3. Secure the material to a table/work bench before biscuit joining.
4. Have a firm grip on the plate jointer with both hands before turning the switch on.
5. Adjust the centering stop so the jointer will cut the biscuit slot in the middle of the edge of the boards to be jointed.
6. Mark the center of all biscuit slot locations before any cuts are made.
7. Align the center notch on the plate jointer with the center mark on the boards to be jointed.
8. Allow the centering stop to rest flush on top of each board to be jointed.
9. Clamp or secure each board before making the cut(s) for biscuit(s).
10. Turn the plate jointer on and push the spring-loaded base in (toward the board). This will expose the blade which will then cut the slot for the biscuit.

11. The plate jointer must be adjusted to cut the correct size slot. There are three different biscuit sizes used for jointery, the plate jointer will have a setting for each size biscuit, adjust accordingly.

GENERAL SAFETY PRACTICES

1. Wear approved eye protection, hearing protection, and proper clothing. Tie up loose hair and remove loose jewelry.
2. Do not operate the machine without the instructor's permission, or without instructor supervision.
3. Ensure the power is off before connecting to an energy source.
4. Disconnect the energy source before making adjustments.
5. Do not lay the plate jointer down until the blade has come to a complete stop.
6. Do not place hands/fingers in line or near the blade.
7. Keep the power cord positioned away from the work to be performed to avoid cutting the cord.

COMPLETION QUESTIONS

1. Position of _______________ should be checked on the plate jointer before the power cord is plugged in?
2. The plate jointer should be laid down after the _______________ has stopped completely?
3. Always wear _______________ when operating the plate jointer?
4. Which of the following should not be worn when operating the plate jointer _______________, _______________?
5. Always hold the plate jointer with _______________ hand(s) when turning on to overcome torque?
6. The biscuit slot should be cut in the _______________ of boards being joined.
7. The center notch on the plate jointer should be aligned with the _______________ on the board(s) to get a correctly positioned biscuit slot(s).
8. There are _______________ different size biscuits available for biscuit jointery.
9. To expose the blade and make a cut you have to _______________ the plate jointer.
10. Place the _______________ away from the blade to prevent it from getting cut.

CIRCULAR SAW

PART IDENTIFICATION

Identify the numbered parts of the circular saw illustrated below.

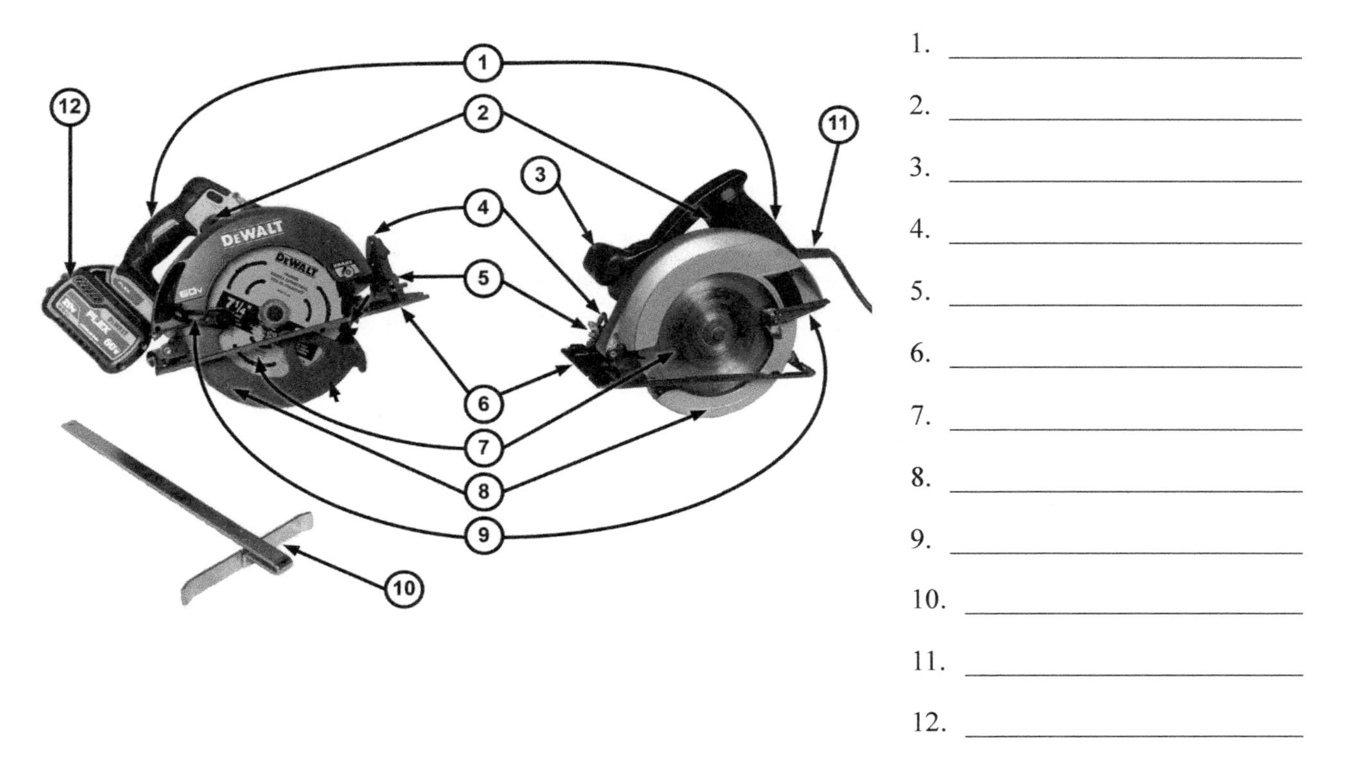

1. ____________________
2. ____________________
3. ____________________
4. ____________________
5. ____________________
6. ____________________
7. ____________________
8. ____________________
9. ____________________
10. ____________________
11. ____________________
12. ____________________

SAFE OPERATIONAL PROCEDURES

1. General Operation
   a. The two basic types of circular saws are the worm drive and sidewinder.
   b. A circular saw can be used for ripping, crosscut, bevels, miters, chamfers, and plunge cuts.
   c. The advertised size of a circular saw is based on the largest blade diameter (i.e. 8¼”, 7¼”, 6½”, 4”) designed for the saw. In general, the most common are the 7¼” and 6½”.
   d. To guide the saw along the cutting line, a notch is provided in the base.
   e. The blade guards consist of a fixed upper portion and a spring-loaded, lower retractable guard that automatically retracts as the saw is moving forward.
   f. Starting the saw is done by squeezing the trigger in the primary handle.
   g. Since most people are right-handed, the blade is positioned on the right side of the motor in most sidewinder saws.
   h. In general, the more teeth on a blade results in a smoother cut.

2. Crosscutting
   a. Use a crosscut or combination blade. Do not change blades or make any adjustment while saw is connected to an energy supply.
   b. Be sure work is solidly supported and secure. Do not cut material between two supports, but allow material to fall or be lifted away from the saw.
   c. Adjust the depth of cut so that the teeth clear the thickness of material by the depth of the teeth.
   d. Grasp primary handle firmly; do not place finger on starting trigger until ready to start saw. Place front of saw base on work so that the guide mark and line of cut are in line.
   e. Advance saw into material, following line of cut. Save the full cutting line by positioning the blade on the waste side of marking line.
   f. Guide the saw steadily through the cut; if you begin cutting off the line, never try to force the saw back on line. Stop the saw and reposition as needed.
   g. When cut is complete, ensure guard closes and blade is stopped before setting the saw down.
3. Ripping:
   a. Use combination or ripping blade.
   b. Use the rip guide attached to the saw or a straight board as a fence to make straight ripping cuts. You may also use a marking line (chalk line) to perform a rip cut.
   c. If saw kerf seems too tight and binds the blade, a wedge can be inserted to open the kerf and give clearance to the blade.
   d. Be sure work is solidly supported and secure. Do not cut material between two supports, but allow material to fall or be lifted away from the saw.
   e. Adjust the depth of cut so that the teeth clear the thickness of material by the depth of the teeth.
   f. Grasp primary handle firmly; do not place finger on starting trigger until ready to start saw. Place front of saw base on work so that the guide mark and line of cut are in line.
   g. Advance saw into material, following line of cut. Save the full cutting line by positioning the blade on the waste side of marking line.
   h. Guide the saw steadily through the cut; if you begin cutting off the line, never try to force the saw back on line. Stop the saw and reposition as needed.
   i. When cut is complete, ensure guard closes and blade is stopped before setting the saw down.
4. Bevel and Miter cuts:
   a. Use combination or crosscut blade.
   b. Lay out angle or line of cut with a sliding T-bevel, speed square, or framing square.
   c. Adjust saw to correct bevel with the angle gauge on the saws if performing a bevel cut.
   d. Adjust blade depth so that blade penetrates bevel thickness of material.
   e. The miter cut is made at an angle other than 90° across the board. The saw blade is set at 90° to the saw frame or zero on the bevel angle scale.

f. A compound bevel cut is a miter cut using a bevel angle.
g. Follow rules as in crosscutting.

5. Plunge cut:
   a. Select a combination blade.
   b. Mark area to be cut.
   c. Adjust the depth of cut so that the teeth clear the thickness of material by the depth of the teeth.
   d. Push lower retractable guard handle forward so lower edge of blade is exposed.
   e. Starting near a corner limit of the pocket to be cut, tilt the saw forward on front of base until the front edge of the blade rests on the surface on the waste side of the line of cut.
   f. With the blade clear of the material, start the saw, lowering slowly into the material until the base rests firmly on material.

GENERAL SAFETY PRACTICES

1. Wear approved eye protection, hearing protection, and proper clothing. Tie up loose hair and remove loose jewelry.
2. Do not operate the machine without the instructor's permission, or without instructor supervision.
3. Use only sharp blades for task to be done.
4. Double check all adjustment thumb screws to be certain they are tight and locked at zero or proper adjustment needed for the task.
5. Be sure the saw is disconnected from energy supply when changing blades or making any adjustments.
6. Never start the circular saw when the blade is in contact with material being cut.
7. Allow the blade to come to a complete stop before setting the saw down.
8. Saw only in forward direction; **never** attempt to saw in reverse.
9. Keep your hand away from the cutting blade during operation.
10. Do not use saw in awkward position such as above head, on a ladder, or on sloping surfaces. Maintain a stable workplace with supports.
11. For a corded saw, be sure saw is properly grounded electrically or double insulated to prevent electrical shock. Determine location of cord at all times to avoid damage to cord. Use a proper extension cord, 3-wire type suitable for outdoor use.
12. Inspect the saw prior to each use for damage-free cord and smooth operating guards.
13. When a blade binds during a cut, stop the saw immediately. This action is referred to as a "kickback." Kickbacks occur when the saw blade is pinched by material being cut and the saw is driven back towards the operator.

14. Allow the saw to reach maximum speed before advancing into the material.
15. Always make sure there is enough slack in the cord so that it can reach the opposite end of the workpiece. If you begin cutting off the line, never try to force the saw back on line. Stop the saw and reposition as needed.
16. When making a cut, place the wide part of the bed/sled on the piece that is not being cut off.
17. Do not talk to anyone while operating the saw.
18. Always check the location of the support material to avoid sawing into it.

COMPLETION QUESTIONS

1. In making a long rip cut, it may be necessary to insert a ________________ in the ______________________ to keep blade from binding.
2. A ______________________ or ______________________ blade can be used for crosscut sawing.
3. A ______________________ saw blade should not be used.
4. In general, the more teeth on a blade results in ______________________ cuts.
5. The ______________________ ______________________ guard must be lifted before making a plunge cut.
6. A ______________________ bevel cut is a miter cut using a bevel angle.
7. The two most common size circular saw are the __________________ and __________________.
8. The saw should be properly ______________________ electrically, to avoid electrical shock.
9. The thickness of the bevel will be ______________________ than the vertical thickness.
10. A ______________________ guide attached to your saw can be used when ripping to assure uniform width of the completed materials.

CUT OFF TOOL

PART IDENTIFICATION

Identify the circled parts on the cut off tool illustrated below.

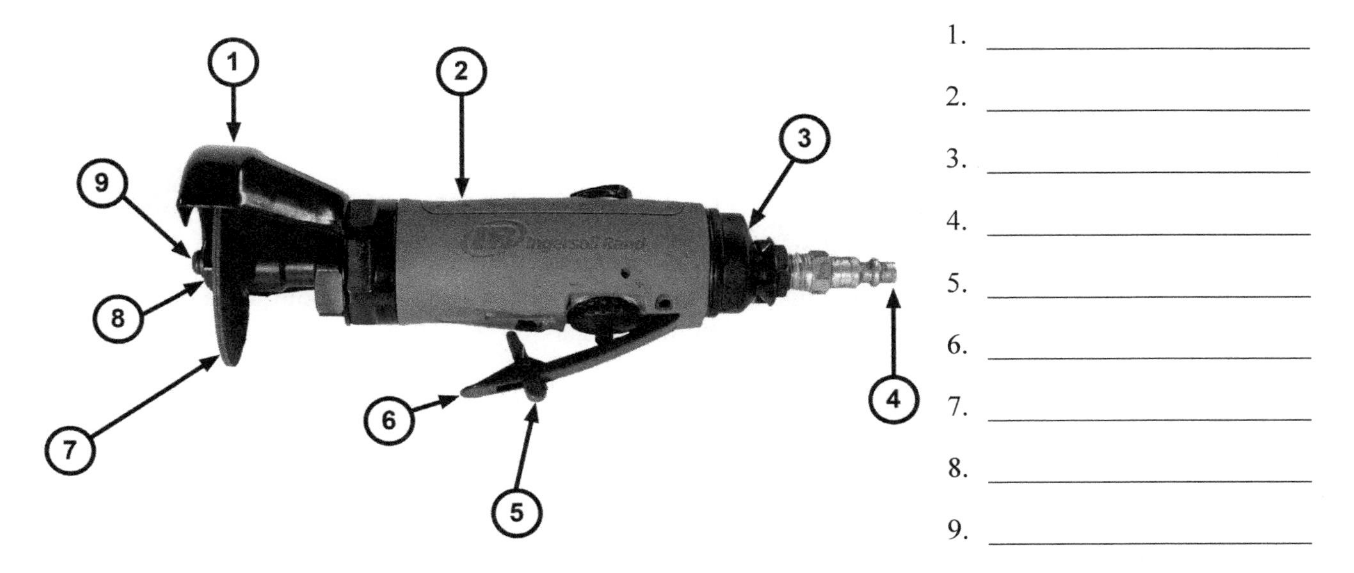

SAFE OPERATIONAL PROCEDURES

1. Lubricate (if pneumatic powered) and clean the tool daily.
2. Make sure the material is secure before using the cut off tool.
3. Periodically stop the tool and check for chips or cracks on the cutting wheel.
4. Run the tool without a load to check the vibration level before using tool. Excessive vibration level is a signal to maintain or repair the tool.
5. Disconnect the power source before replacing the cutting wheel.
6. To install or remove cutting wheel, place open-end wrench on the flats of the spindle below the guard. Keep wrench on spindle stationary while loosening socket head cap screw with hex wrench.
7. Install cutting wheel, making sure directional arrow matches rotation arrow on guard. Replace washer and screw. Tighten securely using the wrenches.
8. With a pneumatic version, blow out air hose to remove any dirt or moisture, then connect the air hose to the tool. Set air pressure to 90 psi.
9. Grip tool with both hands and push safety lever forward to release throttle lever. Gently press throttle lever. Keeping your hands away from the cutting wheel.
10. Allow tool to reach operating speed, then apply moderate pressure to material; do not force tool.

11. Release throttle lever to stop tool. Make sure the cutting wheel has come to a complete stop before setting the tool down.

GENERAL SAFETY PRACTICES

1. Wear approved eye protection, hearing protection, and proper clothing. Tie up loose hair and remove loose jewelry.
2. Do not operate the machine without the instructor's permission, or without instructor supervision.
3. Never remove the cutting wheel guard.
4. Make sure to use the correct cutting wheel for the RPM rating on the tool. Replace the cutting wheel if the cut off tool was dropped and or sustained a heavy impact.
5. The cut off tool throws a large amount of sparks, be sure no one is in the path of the sparks before you begin as well as having proper ventilation.
6. Always check for loose fittings and leaking air hoses on the pneumatic cut off tool.

COMPLETION QUESTIONS

1. _______________ _______________ should be used to protect workers eyes from flying fragments.
2. Remove any loose _______________ when you are cutting.
3. Before you turn on the cut off tool, check the _______________ to be sure it is not cracked.
4. Use _______________ pressure against the wheel, cutting too hard can break the equipment.
5. Never let your _______________ get too close to a cutting wheel.
6. Making sure _______________ arrow on the cutting wheel matches rotation arrow on guard.
7. Excessive _______________ level is a signal to maintain or repair the tool.
8. Tie up your _______________, it could get caught in the spinning wheel.
9. Never remove the cutting wheel _______________.
10. Never exceed _______________ PSI of the cutoff tool.

DIE GRINDER

PART IDENTIFICATION

Identify the circled parts on the die grinder illustrated below.

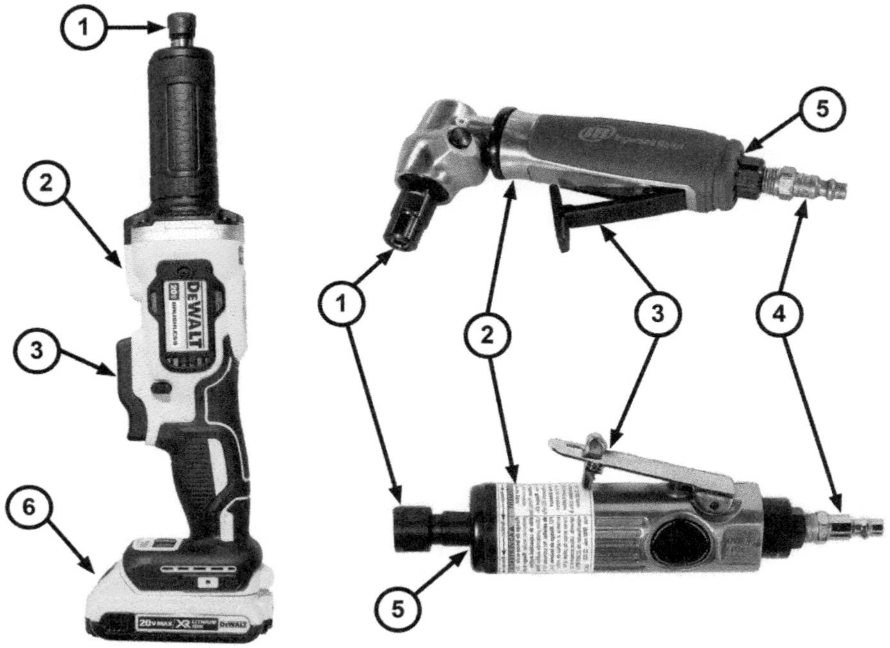

1. ____________________
2. ____________________
3. ____________________
4. ____________________
5. ____________________
6. ____________________

SAFE OPERATIONAL PROCEDURES

1. Lubricate (if pneumatic powered) and clean the tool daily.
2. Secure the material on a table or stand before starting to operate the grinder.
3. If pneumatic powered, set the regulator to the appropriate pressure.
4. Check the grinding stone bit to be sure it is not cracked.
5. Check the grinding burr bit to be sure it is not dull or damaged.
6. Make sure the inserted bit is firmly secured by the collet.
7. Press down the throttle lever to start the tool. The tool will stop running when the throttle lever is released.
8. Check the direction of rotation before use.
9. Run the tool without a load to check the vibration level before using tool. Excessive vibration level is a signal to maintain or repair the tool.
10. Wait until the tool has reached full speed, then gently contact the surface.

11. Periodically stop the tool and check for wear on the grinding bit.
12. Do not use heavy pressure on the tool when operating. Allow the tool to do the work.
13. Move the tool in a uniform pattern up and down or side to side while grinding to ensure even grinding.

GENERAL SAFETY PRACTICES

1. Wear approved eye protection, hearing protection, and proper clothing. Tie up loose hair and remove loose jewelry.
2. Do not operate the machine without the instructor's permission, or without instructor supervision.
3. Wear a face shield when operating a die grinder.
4. Keep your hands away from the grinding bit.
5. To clean the material before working on it.
6. Be sure no one is in the work area.
7. Make sure the work area is clear of flammable items.
8. Hold the tool securely with both hands.
9. Make sure die grinder has come to a complete stop before setting it down.
10. Operate the tool to control dust and fumes, and have proper ventilation.
11. Always check for loose fittings and leaking air hoses on the pneumatic version.

COMPLETION QUESTIONS

1. ________________ ________________ should be used to protect eyes from flying debris.
2. Remove loose ________________ when you are grinding.
3. Before you turn on any grinder, check the ________________ ________________ to be sure it is not cracked.
4. Do not use ________________ pressure against the bit, grinding too hard can break the equipment.
5. Keep your ________________ from getting too close to a grinding bit.
6. Wear a ________________ shield when using a die grinder.
7. Hold the tool ________________ with both hands.
8. Tie up your ________________, it could get caught in the spinning die grinder.
9. Secure the ________________ on a table or stand before starting to operate the grinder.
10. Make sure die grinder has come to a complete ________________ before setting it down.

FINISH SANDERS

PART IDENTIFICATION

Identify the numbered parts of the finish sanders illustrated below.

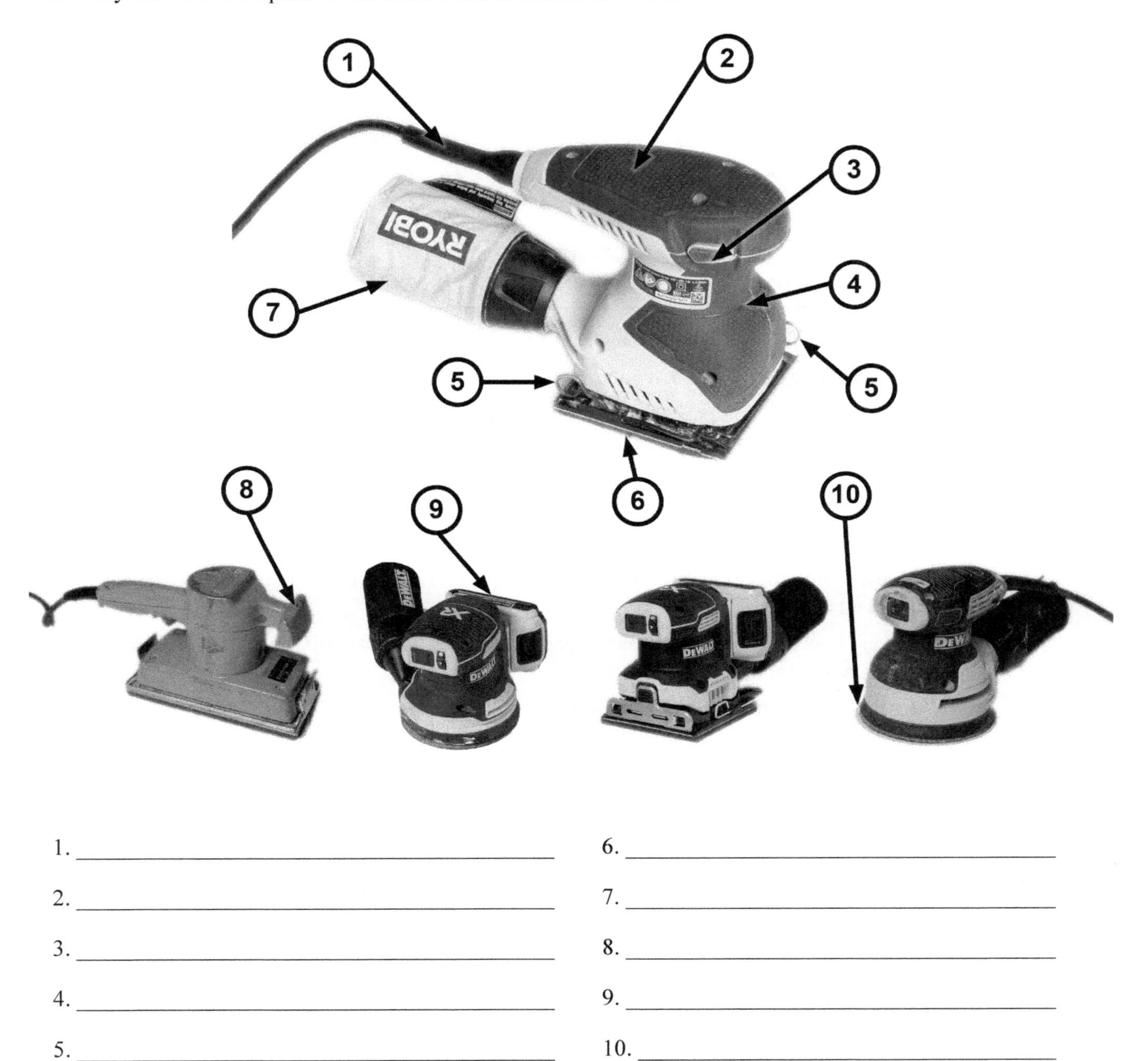

1. ______________________

2. ______________________

3. ______________________

4. ______________________

5. ______________________

6. ______________________

7. ______________________

8. ______________________

9. ______________________

10. ______________________

SAFE OPERATIONAL PROCEDURES

1. Select the proper grit (coarseness or fineness) of sandpaper according to the following table indicating class, grit size of the sandpaper.

CLASS	GRIT SIZE	CLASS	GRIT SIZE
	400		100
	360	Medium	80
Very	320		60
Fine	280		50
	240	Coarse	40
	220		36
	180	Very	30
Fine	150	Coarse	24
	120		20

2. Attaching sandpaper to a sander with a square or rectangular pad:
   a. Make sure energy source is disconnected.
   b. Check condition of sandpaper for rips, tears, or packed with glue or sanding dust.
   c. Select proper grit (coarseness) and size of sandpaper for job. If you are starting with a full sheet, cut the sanding sheets to size or sufficiently crease the sheet before tearing it along a fold to ensure a clean edge.
   d. Open sandpaper clamps.
   e. Insert paper into first clamp, stretch the paper tightly over the pad, and then insert paper into second clamp.
   f. Make sure paper is straight, covers the pad completely, and has sufficient tension so it is tight and snug against pad.
   g. Using the paper punch, punch holes into the sandpaper to allow the dust to flow into the dust bag or vacuum.
3. Attaching sandpaper to a sander with a round pad:
   a. Make sure energy source is removed
   b. Determine if sand paper is attached by adhesive or hook and loop.
   c. Completely remove worn out sandpaper.
   d. Install new paper, ensuring holes are aligned with the holes in the backer pad.
4. Ensure the switch is OFF before connecting energy source.
5. Clamp material to be sanded in vise or to the table.
6. Start the sander above the work. Place sander on workpiece surface, beginning at one side and moving the sander along the length of the material using only enough pressure (normal weight of the machine) to sand. Always sand with the wood grain. Sanding is generally done with one hand.

7. If the sander doesn't seem to be sanding fast enough, switch to a coarser grit. Do not apply more pressure.
8. Use successively finer grit paper until desired finish is obtained.
9. When sanding is completed, lift the sander off the work before stopping the motor. Wait for the sander to stop; then set the sander on the table.
10. Before storing the sander, thoroughly clean dust and empty the dust bag.

GENERAL SAFETY PRACTICES

1. Wear approved eye protection, hearing protection, and proper clothing. Tie up loose hair and remove loose jewelry.
2. Do not operate the machine without the instructor's permission, or without instructor supervision.
3. Never operate machine with torn or worn sandpaper.
4. Make sure sander switch is off before connecting to energy source.
5. Check to see that paper is secure before operating machine.
6. Do not overload machine with excessive downward pressure on material.
7. Keep machine clean.

COMPLETION QUESTIONS

1. Ensure the holes in the sandpaper line up with the holes in the __________________ when installing new sandpaper.
2. Ensure the switch is __________________ prior to connecting to an energy source.
3. Start sanding with a coarse grit working toward a __________________ grit.
4. A number 80 grit size would be considered __________________ class sandpaper.
5. Downward pressure on the material is generally enough just from the __________________ of the machine.
6. The paper should be inserted into the __________________ and then stretched over the __________________ and locked into the second clamp.
7. Sandpaper should be __________________ to the pad.
8. The machine should be moved __________________ the wood grain.
9. The sanding operation is normally completed by holding the machine with __________________ hand.
10. The use of unevenly torn paper could result in damage to the sander __________________ so sufficiently crease the paper along a fold before tearing to ensure a clean edge.

GAS METAL ARC WELDER
(MIG WELDER OR WIREFEED WELDER)

PART IDENTIFICATION

Identify the circled parts on the gas metal arc welder (MIG welder or wire feed welder) illustrated below.

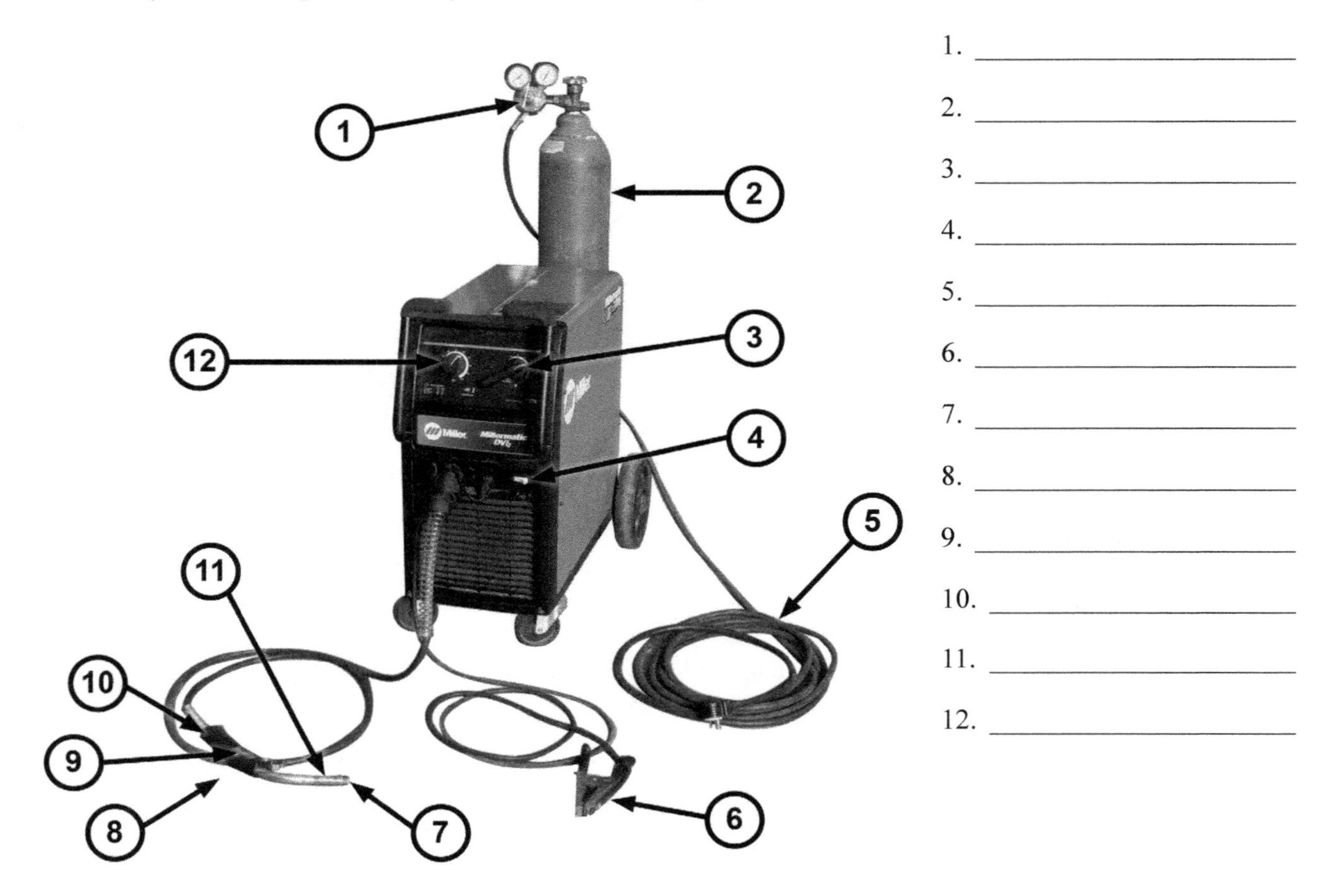

SAFE OPERATIONAL PROCEDURES

1. Check to make sure the welding machine is properly grounded. The welding equipment should be installed according to provision of the National Electric Code and the manufacturers recommendations.
2. A power disconnect switch should be conveniently located near each welding machine.
3. The operator should keep all cable connections tight and inspect all cables, cords and gun parts prior to welding to be sure everything is working properly.
4. Be sure you are wearing ALL the proper welding attire before turning on the welder.

5. Be sure you are in an approved welding area equipped with proper ventilation and shielded from others to stop them from getting arc flash.
6. Turn on the fumes removal system before starting to weld.
7. Be sure that all gas cylinders are chained in an upright position before starting to weld.
8. Clear all combustible materials from the welding area before welding.
9. Use an audible signal such as "cover" or "welding" to indicate to others that you plan to pull the trigger.
10. Check to be sure the gun is not clogged before welding. Too much spatter can prevent the shielding gas from coming out and shielding your weld bead. Use needle-nose pliers to clean the tip; never pound the tip on the bench or floor.
11. Check your machine settings and set them to the manufacture's recommendations. There is usually a chart on the inside of the welder's hood that tells you what the wire feed speed and voltage should be set to depending on the type of gas and wire you are using and thickness of metal you are welding.
12. When opening the regulator for your gas, never stand directly in front of them in case the pressure build up causes the regulator to explode. Crack it a bit at first and then open it all the way to make sure the seat inside sits properly and you will get proper gas flow.
13. Turn the regulator screw in a clockwise position until it reads between 15-20 CFH, depending on the job you are doing. Use lower gas flows for practice and higher ones for actual projects.
14. Be sure that you are in a comfortable position and that you can visually see what you are doing. Make a dry run before pulling the trigger. This will help ensure that you will be able to weld your entire weld bead efficiently and effectively.
15. Before pulling the trigger, for a flat bead, align your gun so it is at 90-degree angle to you and a 15-20 degree angle in the direction of travel. Pull the trigger and use a steady motion when welding. The rate of travel of the welding tip will depend on the weld being done.
16. Cool the metal after each weld so that the metal does not get too hot, especially with thinner metal pieces.
17. If a small ball of metal has formed on the end of the wire during the welding process, cut it off with the wire cutters so about 1/8" sticks out of the tip. This must be done often so the electricity can connect from the metal wire to the piece of metal more easily and makes for a nicer looking weld start.
18. If the metal wire melts to the tip, tell the instructor so a new tip can be put in its place.
19. Hang up MIG gun and turn off welder when work is being changed or when work has been completed. Be sure to turn the welder off and store all cables before leaving the welding area.

GENERAL SAFETY PRACTICES

1. Wear approved eye protection, hearing protection, and proper clothing. Tie up loose hair and remove loose jewelry.

2. Do not operate the machine without the instructor's permission, or without instructor supervision.
3. Use an approved helmet with minimum #10 shaded lens for non-ferrous and #12 for ferrous metals.
4. Always wear protective clothing suitable for welding. Wool or cotton clothing, close-toed shoes (high top leather shoes recommended), leather gauntlet gloves, welding coat or leathers. DO NOT wear clothing made of synthetic fibers when welding. Some synthetic fibers are highly flammable. All this must be worn to prevent burns from ultraviolet and infrared rays emitted while arc welding.
5. Guard against the use of damp or wet clothing when welding. The use of such clothing increases the possibility of electrical shock.
6. Never weld while standing in water or on damp ground. Dampness on the body increases the chance of electrical shock when welding.
7. Do not talk with observers while operating machines. Give the welder your full attention.
8. Do not carry matches, butane or propane lighters or other flammables in your pockets while welding.
9. Never breath fumes when welding lead, cadmium, chromium, steel, manganese, brass, bronze, beryllium, zinc, or galvanized steel. These fumes are toxic and may cause sickness or death. A good exhaust system is essential when welding within a confined laboratory.
10. Shield others from the light rays produced by MIG welding. Keep the welding curtain in place at all times to protect others from arc flash.
11. Keep the welding area clean and free of tools, scrap metal, and water.
12. Make sure the work area is free of flammable, volatile, or explosive materials. (Ex. propane, gasoline, grease, paper and coal dust).
13. Protect welding cables from sparks, hot metal, open flames, sharp edges, oil, and grease. Do not use cables with frayed, cracked or bare spots in the insulation.
14. Never weld with the cables coiled over the shoulders.
15. Never lay the MIG gun on the welding table or a grounded metal surface in case the trigger is accidentally depressed. Place it on an insulated hanger.
16. Report to supervisor at once if electrode holder, cable connection, cable, or cable terminals at the welding machine, ground clamps or lugs get hot.
17. Never touch the MIG wire while the welder is turned on. It is electrically "hot" and can cause a serious shock.
18. Handle all compressed gas cylinders used in MIG welding with extreme care. Keep the cylinder caps in place when the cylinders are not in use. Do not weld in areas that store compressed gas cylinders.
19. When gas cylinders are empty, close the valve and mark cylinders "empty."

20. Use tongs or pliers to handle hot metal after it has been welded. If quenching, completely submerge metal in water when cooling, this prevents steam from burning you. Cool and store any hot metal before leaving the work area.

21. Avoid welding directly on concrete floors. Residual moisture in the concrete may be turned to steam resulting in the concrete exploding.

22. Disconnect the power to a welding machine before making any repairs.

23. Use a fire blanket to smother clothing fires. Use a dry chemical type "C" extinguisher to put out an electrical fire.

COMPLETION QUESTIONS

1. Check to make sure the welding machine is properly ________________.

2. Be sure you are in an approved welding area equipped with proper ________________.

3. Be sure that all ________________ ________________ are chained in an upright position before starting to weld.

4. When opening the regulator for your gas, never stand directly in ________________ of them in case the pressure build up causes the regulator to explode.

5. ________________ ________________ MIG gun and turn off welder when work is being changed or when work has been completed.

6. Use an approved helmet with minimum ________________ shaded lens for non-ferrous metals.

7. DO NOT wear clothing made of ________________ fibers when welding.

8. Do not carry ________________, butane or propane lighters or other flammables in your pockets while welding.

9. When gas cylinders are empty, close the valve and mark cylinders "________________."

10. Use tongs or ________________ to handle hot metal after it has been welded.

GAS TUNGSTEN ARC WELDER (TIG WELDER)

PART IDENTIFICATION

Identify the circled parts on the gas tungsten arc welder (TIG welder) illustrated below.

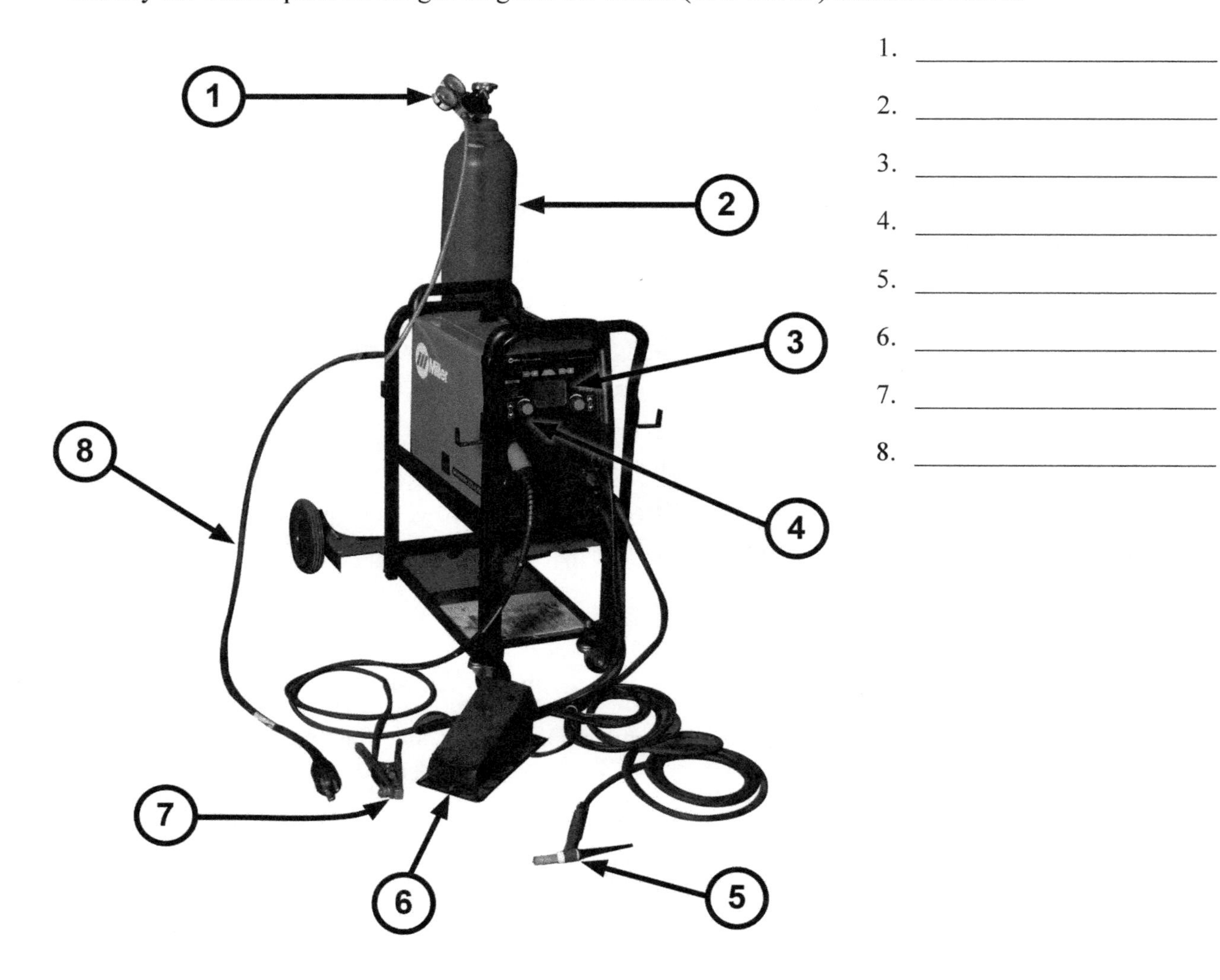

1. ____________________
2. ____________________
3. ____________________
4. ____________________
5. ____________________
6. ____________________
7. ____________________
8. ____________________

SAFE OPERATIONAL PROCEDURES

1. Have work area properly ventilated. The ventilation system must be turned on prior to welding.
2. Remove all flammable material from the welding work area.
3. Don’t touch the tungsten electrode to your skin when the welder is on.
4. Select the proper type and diameter of tungsten electrode.

5. Select the shape of the tip. Mild steel welding needs a point and aluminum welding needs a ball at end of the tungsten rod.
6. Set the type of current, current range, and current amount.
7. Set the high-frequency switch to the desired position if applicable.
8. Set the post flow timer for desired post flow.
9. Make sure shielding gas is on during operations
10. Attach the ground clamp as close to the where you will be laying a bead as possible. The shorter the distance electricity has to travel to ground the safer the operator will be.
11. Insulate yourself from the work bench, this will prevent any shock.
12. Do not touch hot material, use pliers or suitable tool for handling hot material. Protect hot material from contact by others.
13. TURN OFF shielding gas after welding job is complete.
14. Turn off welding power source after the unit has cooled down.

GENERAL SAFETY PRACTICES

1. Wear approved eye protection, hearing protection, and proper clothing. Tie up loose hair and remove loose jewelry.
2. Do not operate the machine without the instructor's permission, or without instructor supervision.
3. Additional protective welding clothing, including a helmet, long-sleeve jacket, and long cuff leather gloves, must be worn to prevent burns from ultraviolet and infrared rays emitted while welding.
4. The helmet used for TIG welding should be equipped with a minimum number ten density shade. A welding helmet is NOT approved for eye protection, the shaded lens is not impact resistant, therefore eye protection still must be worn under the helmet.
5. Any dangerous condition, faulty equipment, or injury no matter how small such as a burn, must be reported to the instructor.
6. Be careful moving and working around gas cylinders.
7. Prevent exposure to dangerous fumes from welding. For example, never weld wet aluminum and only a professional should weld galvanized material.
8. Keep work clean and clear from being cluttered with tools and material.
9. Welders can be electrocuted, stay clear from water or wet materials in the work area.
10. An electric shock can occur, keep your equipment and material grounded when welding.
11. Do not coil or drape cables around your body.

12. Slag and sparks occur when welding, be aware of possible fires and explosions.
13. Use a safety shield to protect people in surrounding area from eye damage.
14. Turn off welding power source before changing tungsten electrode or torch parts.

COMPLETION QUESTIONS

1. Eye protection is ______________ to be worn under the welding helmet.
2. Any dangerous condition or injury must be reported to the ______________.
3. Make sure shielding gas is ______________ during operations and TURN OFF after welding job is complete.
4. If a piece of equipment is not working properly, tell your ______________.
5. The ventilation system must be turned on ______________ to welding.
6. Wear leather gloves, this will prevent ______________.
7. Remove all ______________ material from the welding work area.
8. Be sure to ______________ your material to the table to ensure you have a good ground.
9. The helmet used for TIG welding should be equipped with a minimum number ______________ density shade.
10. Welders can be ______________, stay clear from water or wet materials in the work area.

HEAT GUN

PART IDENTIFICATION

Identify the circled parts on the heat gun illustrated below.

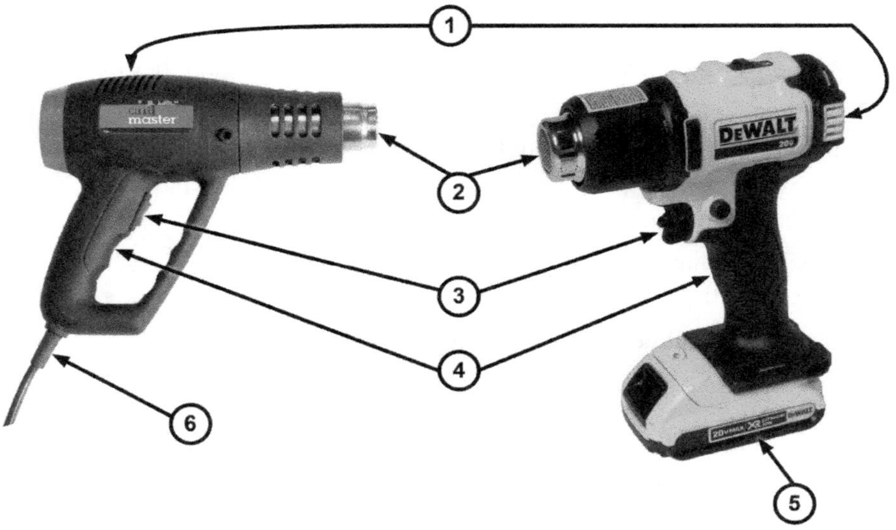

1. ____________________
2. ____________________
3. ____________________
4. ____________________
5. ____________________
6. ____________________

SAFE OPERATIONAL PROCEDURES

1. General Heat Gun Use
   a. Prepare your work area by removing flammable materials.
   b. Prepare heat resistant surfaces for the heat gun and heated components.
   c. Read operators manual to determine the heat settings needed for the operation.
   d. Connect the heat gun to an energy source.
   e. Heat the area needed slowly. Too much concentrated heat can cause items to warp and break.
   f. Apply just enough heat to complete the task.
   g. Rest the heat gun on its stand to allow to cool down.
   h. Cool all parts before proceeding.
   i. Unplug the tool when complete, and let it cool completely before storing.
2. Removal of Finishes (non-lead based)
   a. Move the work piece outdoors. If this is not possible, keep the work area well ventilated. Open the windows and put an exhaust fan in one of them. Be sure the fan is moving the air from inside to outside.
   b. Remove or cover any carpets, rugs, furniture, cooking utensils, and air ducts.
   c. Place drop cloths in the work area to catch any paint chips or peelings. Wear protective clothing such as extra work shirts, overalls and hats.

d. Work in one room at a time. Furnishings should be removed or placed in the center of the room and covered. Work areas should be sealed off from the rest of the dwelling by sealing doorways with drop cloths.

e. Wear a dust respirator mask or a dual filter (dust and fume) respirator mask which has been approved by the Occupational Safety and Health Administration (OSHA) or the National Institute of Safety and Health (NIOSH).

f. Use caution when operating the heat gun. Keep the heat gun moving as excessive heat will generate fumes which can be inhaled by the operator.

g. Keep food and drink out of the work area. Wash all exposed skin and rinse mouth before eating or drinking.

h. Clean up all removed paint and dust by wet mopping the floors. Use a wet cloth to clean all walls, sills and any other surface where paint or dust is clinging. Do not sweep, dry dust or vacuum.

i. At the end of each work session put the paint chips and debris in a double plastic bag, close it with tape or twist ties, and dispose of properly.

j. Remove protective clothing and work shoes in the work area to avoid carrying dust into the rest of the dwelling. Wash work clothes separately. Wipe shoes off with a wet rag that is then washed with the work clothes.

GENERAL SAFETY PRACTICES

1. Wear approved eye protection, hearing protection, and proper clothing. Tie up loose hair and remove loose jewelry.
2. Do not operate the machine without the instructor's permission, or without instructor supervision.
3. If possible, operate the heat gun outdoors.
4. If operating indoors, provide adequate ventilation.
5. The heat gun will get hot. Only rest the heat gun on the stand away from flammable materials.
6. The forced air coming from the nozzle is hot enough to burn your skin. Never direct the air towards another person.
7. The metal nozzle will become hot during use. Do not touch the hot nozzle.

COMPLETION QUESTIONS

1. Prepare your work area by removing ________________materials.
2. Heat the area needed ________________. Too much concentrated heat can cause items to warp and break.
3. Rest the heat gun on its ________________ to allow to cool down.
4. When removing finishes, move the workpiece ______________ if possible.

5. Remove or ______________any carpets, rugs, furniture, cooking utensils, and air ducts.
6. Wear a dust respirator mask or a dual filter (dust and fume) respirator mask which has been approved by______________ or ______________.
7. Keep food and drink out of the work area. Wash all exposed ______________ and rinse mouth before eating or drinking.
8. If operating indoors, provide adequate ______________.
9. The forced air coming from the nozzle is hot enough to ______________ your skin. Never direct the air towards another person.
10. The metal ______________ will become hot during use. Do not touch it.

HEAT PRESS

PART IDENTIFICATION

Identify the circled parts on the heat press illustrated below.

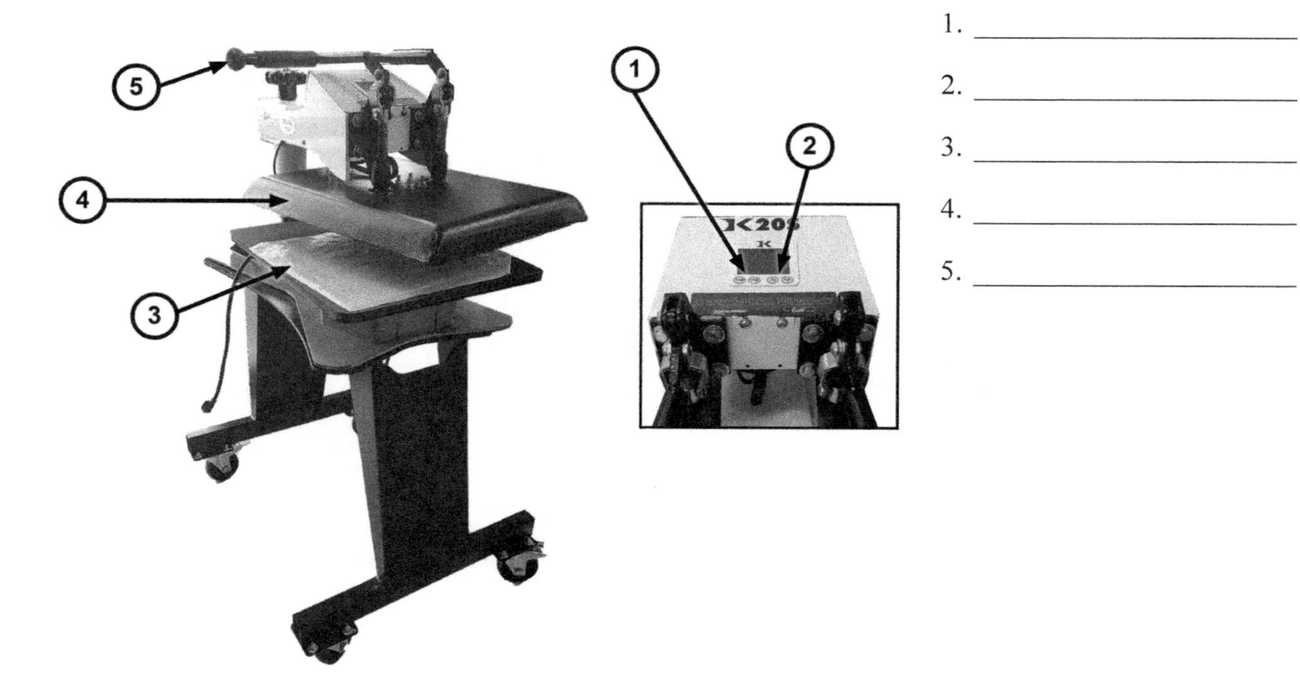

SAFE OPERATIONAL PROCEDURES

1. Connect the heat press to power socket using the connection cable supplied.
2. Switch on the press with the red rocker switch.
3. After switching on, the press heats up automatically to the set temperature, which can be changed at any time on the temperature control knob.
    a. A small light on the temperature control knob illuminates as long as the press is heating up.
4. You can adjust the pressure by increasing the pressure arm of the lever.
    a. To increase or decrease the pressure, turn the round hand knob on top of the heat press.
5. As soon as the press is heated up completely you can start making the first transfer.
6. For this purpose, select the desired pressing time on the time control knob. The timer starts running as soon as the press is closed with the press lever.
7. After expiration of the set time a warning signal sounds.

8. Maintenance and cleaning maintenance work should always be accomplished with the press switched off and cooled down.
9. Disconnect the plug from the energy source before starting work.
10. Clean the press regularly with a soft cloth and mild household cleaner to remove adhesive residues, etc.
11. Never use scouring sponges, solvents or gasoline!

GENERAL SAFETY PRACTICES

1. Wear approved eye protection, hearing protection, and proper clothing. Tie up loose hair and remove loose jewelry.
2. Do not operate the machine without the instructor's permission, or without instructor supervision.
3. Never reach into the heat press with your hands when it is connected to the energy source, particularly when it is switched on and heated up. There is a danger of severe burns!
4. Never open the housing or attempt to modify the machine yourself.
5. Ensure that liquids and metal objects do not get into the inside of the heat press.
6. Ensure that the power socket used is grounded. Note that it is only permissible to operate a heat press from a power socket protected by a ground fault protection switch.
7. Disconnect the power plug if the machine is not to be used for a longer period of time.
8. Never operate the heat press within the reach of children and never leave the machine unsupervised when switched on.
9. Ensure that the machine is used only in dry rooms.

COMPLETION QUESTIONS

1. After switching on, the press heats up ________________ to the set temperature, which can be changed at any time on the temperature control knob.
2. You can adjust the ________________ by increasing the pressure arm of the lever.
3. It may take some ________________ for the press to cool down after it has been switched off.
4. Maintenance and cleaning maintenance work should always be ________________ with the press switched off and cooled down.
5. Clean the press ________________ with a soft cloth and mild household cleaner to remove adhesive residues, etc.
6. Never use scouring sponges, ________________ or gasoline!
7. Never ________________ ________________ the heat press with your hands when it is connected to the power supply.

8. Never open the ________________ or ________________ to modify the machine yourself.
9. Ensure that ________________ & ________________ objects do not get into the inside of the heat press.
10. Never operate the heat press within the ________________ of ________________ and never leave the machine unsupervised when switched on.

HOT GLUE GUN

PART IDENTIFICATION

Identify the circled parts on the hot glue gun illustrated below.

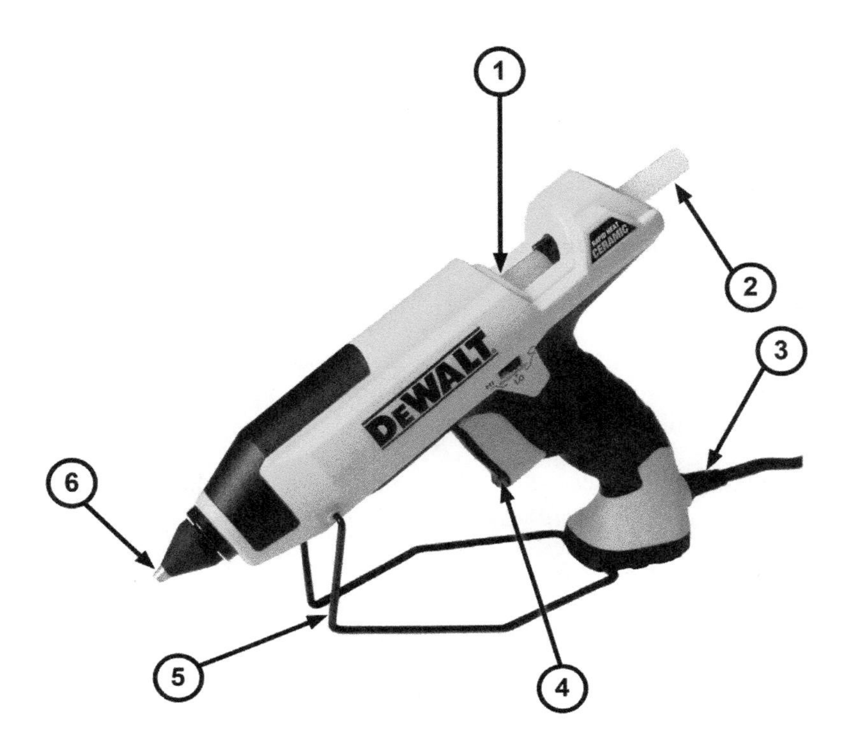

1. ____________________
2. ____________________
3. ____________________
4. ____________________
5. ____________________
6. ____________________

SAFE OPERATIONAL PROCEDURES

1. Choose the kind of gun you need for your project. If you don't need a high heat gun, opt for one of the "warm" or "low melt" options to reduce (but not eliminate) your chances of injury. Glue guns called "warm" or "low melt" guns can still be dangerous and result in burns or electrical shocks.
2. If you're using a dual-temp glue gun, you can change the temperature of the glue depending on what materials you're using. If you need a stronger bond with materials like ceramics, leather, metal or wood, use a higher heat, but if you're using more fragile materials like paper, flimsy fabric or lace, leave it on a lower heat setting.
3. Before you plug in the hot glue gun, carefully inspect it. Check to make sure the gun isn't broken or too worn. Look for a frayed electrical cord or any cracks in the gun or nozzle. If you find these signs of excessive wear, do not use the glue gun.
4. Plug the gun in and check to see if the on/off button (if your gun has one) is working correctly.
5. Wait for it to heat up before squeezing glue out of the gun.
6. When it is warm enough, the trigger should be able to be depressed with ease.

7. Do not tilt a hot glue-gun nozzle upwards or attempt to use a hot glue gun to glue overhead items.
8. When not using the gun, make sure to set it down upright on its metal rack, instead of lying it on its side.
9. Place something underneath the tip like tin foil or a piece of ceramic tile to catch any glue that drips out of the hot glue gun tip.
10. Once complete, unplug the hot glue gun and let it cool before wrapping up the cord and putting it away.

GENERAL SAFETY PRACTICES

1. Wear approved eye protection. Tie up loose hair and remove loose jewelry.
2. Do not operate the glue gun without the instructor's permission, or without instructor supervision.
3. Do not touch the hot nozzle or hot glue when working with your glue gun. If you accidentally do and burn yourself, tell your instructor immediately and cool the burn under cold water.
4. Do not leave your glue gun plugged in and unattended—this can be a fire hazard and be dangerous for other people who may come in contact with the hot appliance.
5. Only use glue sticks that are recommended for your particular glue gun.
6. Prevent inhalation of hot-glue vapors by using the glue gun in well ventilated areas.
7. Keep your glue gun out of direct sunlight or any moist conditions to reduce the risk of electrical shock or fire.

COMPLETION QUESTIONS

1. Low temperature glue guns will ________________ the chance for injury.
2. Ensure the ________________ is set correctly according to the material you are gluing.
3. Depress the ________________ with ease. Don't use excessive force.
4. Do not tilt a hot glue-gun nozzle ________________ or attempt to use a hot glue gun to glue overhead items.
5. Use only the recommended ________________ ________________ for your glue gun.
6. Do not leave your glue gun plugged in and ________________.
7. When not using the gun, make sure to set it down ________________ on its metal rack, instead of lying it on its side.
8. Place something underneath the ________________ to catch any glue that drips out.
9. Only use glue sticks that are ________________ for your particular glue gun.
10. Prevent inhalation of hot-glue vapors by using the glue gun in well ________________ areas.

IMPACT WRENCH

PART IDENTIFICATION

Identify the numbered parts of the impact wrench illustrated below.

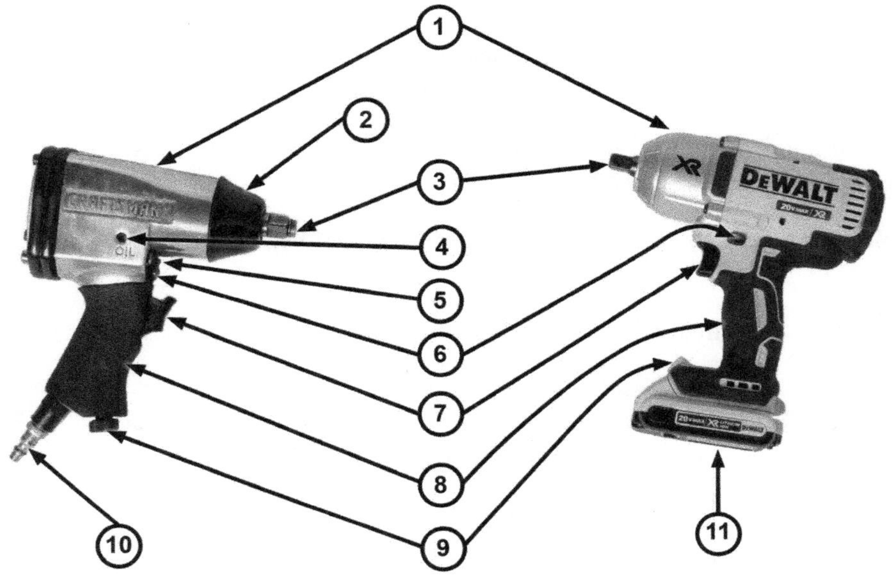

1. ____________________
2. ____________________
3. ____________________
4. ____________________
5. ____________________
6. ____________________
7. ____________________
8. ____________________
9. ____________________
10. ____________________
11. ____________________

SAFE OPERATIONAL PROCEDURES

1. Study the operator's manual that accompanies the specific model and make of the impact wrench.
2. The impact wrench is powered by regulated air and is designed to be used to loosen and tighten nuts, bolts, and screws to a predesigned torque value.
3. Before opening the main air valve, check all air hoses, pipe fittings, pressure gauges, and regulators for leaks or damage.
4. Operating air pressure should be no lower than 70 PSI or higher than 120 PSI with 90 PSI being the recommended air hose pressure.
5. Check the CFM (cubic feet/minute) volume of the air supply. Most impact wrenches require 3 to 5 CFM for proper operation.
6. Check the air inlet plug (male connector) and air coupler (female connector) to make sure the plug and coupler match.
7. Select correct impact-only drive socket for job to be completed. Do not use regular sockets as they are not designed for impact wrench application. Impact sockets are generally black, have a thick wall, and are manufactured for impact wrench use.

8. Make all adjustments on wrench with the air hose disconnected. Adjust the output torque control for desired bolt torque. Adjust the reversing valve for tightening or loosening fasteners.
9. Lubricate the impact tool according to manufacturer's recommendation. If an air hose oil lubricator is used, check for oil supply and proper operation. If oil is added into air hose, direct lubrication of the tool is not necessary.
10. When the impact wrench is properly adjusted and lubricated, connect to the air hose.
11. Impact wrenches should not be operated in more than a quick burst without contact to the fastener. Operation without contact to fastener could cause the socket to fly from the drive.
12. Make sure the material containing fasteners to be tightened or loosened is secured in a vise or held by appropriate means, never by hand or by a helper.
13. When tightening or loosening fasteners, use the tool in short bursts to avoid stripping or as a check for the correct rotation.
14. If more or less tightening torque is required, adjust the output torque control screw accordingly.
15. When work is completed, shut off the air supply to the air hose, remove any attached impact sockets, and vent compressed air in air hose by squeezing the trigger.
16. Disconnect the impact wrench from the air hose, remove impact socket, clean the tool, and place in proper storage.

GENERAL SAFETY PRACTICES

1. Wear approved eye protection, hearing protection, and proper clothing. Tie up loose hair and remove loose jewelry.
2. Do not operate the machine without the instructor's permission, or without instructor supervision.
3. Make all adjustments, lubricate, and clean the tool before connecting to the air hose.
4. Disconnect the wrench from the air hose when not using or when leaving the work area.
5. Carry the tool only by the handle and not by the air supply hose.
6. Select impact sockets that are designed for the type of work to be completed.
7. Make sure air connectors (plug and coupling) are matched.
8. Use only regulated air; never use bottled air or gases to power the impact wrench.
9. Use only recommended air pressure; under-pressure operation may be as dangerous as over-pressure operation.
10. Make sure the material is secured before attempting to loosen or tighten fasteners.
11. Operate the impact wrench only in short bursts and only when making contact with fasteners.

12. Shut off the air supply and bleed the air hose by squeezing the trigger before disconnecting the tool from the air hose.

COMPLETION QUESTIONS

1. The recommended air pressure for operating the impact wrench is ________________ PSI.
2. The air supply volume in CFM for most impact wrenches is ________________________ to ______________________ CFM.
3. The impact wrench should be carried by the ________________________ and not by the air ________________________.
4. The ________________________ valve on the wrench is adjusted to change rotation.
5. The ________________________ ________________________ control is adjusted to obtain the correct tightening of fasteners.
6. The impact wrench should be ________________________ daily unless oil is added directly into the air supply line.
7. The impact wrench should only be operated with ________________________ air and not on bottled air or gases.
8. Specially designed impact ________________________ should be used as regular sockets have ________________________ walls which could be damaged during use and cause injury.
9. The air wrench should only be operated for short ________________________ and only when in ________________________ with fastener.
10. Only the ________________________ plug connector should be connected to the tool while the ________________________ coupler should be connected to the air hose.

JACKS

PART IDENTIFICATION

Identify the circled parts on the jacks illustrated below.

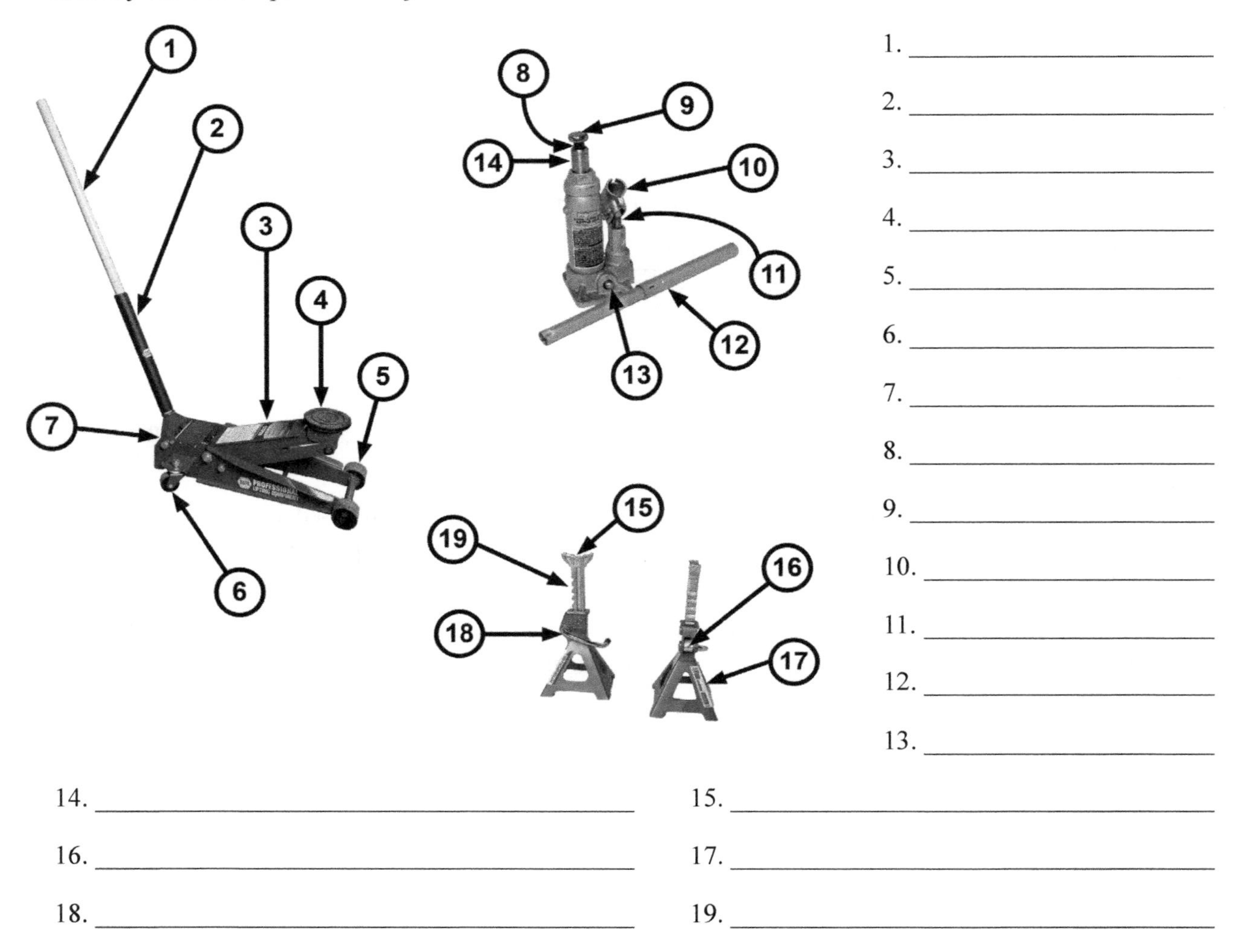

1. ______________________
2. ______________________
3. ______________________
4. ______________________
5. ______________________
6. ______________________
7. ______________________
8. ______________________
9. ______________________
10. ______________________
11. ______________________
12. ______________________
13. ______________________

14. ______________________ 15. ______________________

16. ______________________ 17. ______________________

18. ______________________ 19. ______________________

SAFE OPERATIONAL PROCEDURES

1. Raising the Jack
   a. Place vehicle in park, with emergency brake on and wheels securely chocked to prevent inadvertent vehicle movement.
   b. Locate and close release valve by turning pump handle clockwise until firm resistance is felt.
   c. Verify lift point, center jack saddle under lift point.
   d. Insert pump handle into handle sleeve and pump to contact lift point. To lift, continue pumping until load reaches desired height.

e. Transfer the lifted load to a matched pair of appropriately rated jack stands.

2. Installing Jack Stands
   a. Adjust height by pulling up on ratchet bar.
   b. Locking handle weight should secure the ratchet bar in desired position. To confirm this, push down on the locking handle. Ensure locking handle and ratchet bar are secure before loading.
   c. Carefully position jack stands so that load is centered on saddle.
   d. Slowly lower the vehicle onto the stands.
   e. Ensure vehicle is secure before working on, around or under. Chock all wheels remaining on the ground in both directions to prevent inadvertent movement.
3. Removing Jack Stands
   a. Using suitable jack, raise vehicle clear of stands.
   b. Release locking handle and adjust ratchet bar to lowest position.
   c. Remove stands, then lower vehicle.
4. Lowering the Jack
   a. Raise load high enough to clear jack stands; then carefully remove them.
   b. Slowly turn handle counter-clockwise, but no more than 1/2 turn. If load fails to lower:
      (1) Use another jack to raise vehicle high enough to reinstall jack stands.
      (2) Remove malfunctioning jack, then repeat lowering procedure from step a.
      (3) Using functioning jack, lower vehicle.
   c. After removing jack from under vehicle, push saddle down to reduce ram exposure to rust and contamination.

GENERAL SAFETY PRACTICES

1. Wear approved eye protection, hearing protection, and proper clothing. Tie up loose hair and remove loose jewelry.
2. Do not operate the machine without the instructor's permission, or without instructor supervision.
3. Check your jack for obvious signs of wear and tear, in particular watch for any fluid leaks.
4. Look for any obvious structural damage, in particular in the joints of the jack.
5. Inspect stands before each use. Do not use if bent, broken or cracked components are noted. Ensure that all parts move freely.
6. Check that your jack is rated to take the weight of your vehicle.
7. Ensure the vehicle is in park and the emergency brake is on before beginning to raise your vehicle.
8. Use wheel chocks to prevent the car from rolling.

9. Place the jack on a flat, stable surface. Level surfaces such as bitumen or concrete are best.
10. A jack is a lifting device only. Immediately after lifting, support the vehicle with appropriate means.
11. Lift only on areas of the vehicle as specified by the vehicle manufacturer.
12. Take your time. Don't rush the process. This applies more so when releasing the pressure in the jack to lower your vehicle.
13. Test stability before reaching under the vehicle.
14. Do not move or dolly the vehicle on the jack.
15. No alteration shall be made to the jack.
16. Center load on saddle.

COMPLETION QUESTIONS

1. Check that your ___________________ is rated to take the weight of your vehicle.
2. Center the load on ___________________ of the jack.
3. Look for any obvious ___________________ ___________________, in particular in the joints of the jack.
4. Use wheel ___________________ to prevent the vehicle from rolling
5. Verify ___________________ point, center jack saddle under lift point.
6. Place the jack on a flat, ___________________ surface.
7. Test ___________________ before reaching under the vehicle.
8. ___________________ lower the vehicle onto the jack stands.
9. No ___________________ shall be made to the jack.
10. Take your ___________________. Don't rush the process.

JIG SAW

PART IDENTIFICATION

Identify the numbered parts of the jig saw illustrated below.

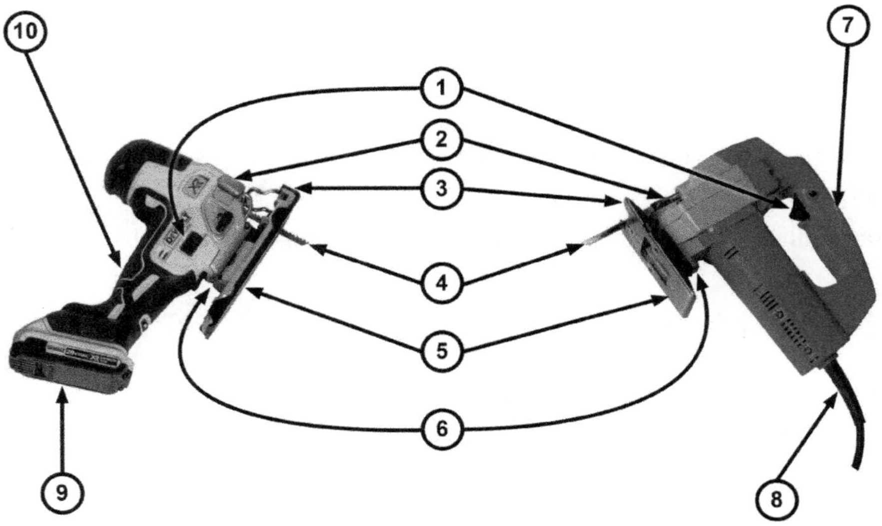

1. ____________________
2. ____________________
3. ____________________
4. ____________________
5. ____________________
6. ____________________
7. ____________________
8. ____________________
9. ____________________
10. ____________________

SAFE OPERATIONAL PROCEDURES

1. Replacing the jig saw blade:
   a. Many blades are available for use in the jig saw. Be sure the blade you use is the correct type for the material being cut and process being performed. There are blades for wood cutting, metal cutting, and knife blades used for cutting soft materials like cardboard. There are also many blades available within each blade type. Follow manufacturer's blade recommendations for each job. Note: Two (2) teeth or more should be in contact with the cutting surface at a time.
   b. Be sure the blade has the correct shank. One blade shank will not fit all brands of saws or even all models of saws within one brand.
   c. To remove blade, loosen blade screw OR extend the quick release lever depending on design and then pull the blade from the slot. (Turn the blade screw back in several turns to prevent loss if applicable.)
   d. To place a blade in the saw, extend the quick release lever OR loosen blade screw depending on design and insert blade shank until seated or hole in shank is aligned with blade screw. Teeth must be facing forward and pointing upward; then allow lever to close OR tighten screw firmly depending on design.
   e. Be sure the saw is disconnected from the energy supply before making adjustments or changing blades.

2. Adjustment of the jig saw:
   a. Many jig saws have bases which may be tilted for bevel cuts. These saws characteristically have an arc divided in 5, 10, or 15-degree increments located to the front or rear of the aluminum housing.
   b. For beveled cuts, loosen the bevel locking screw, tilt the base until the desired angle is indicated on the arc, tighten the bevel locking screw, and check the arc and pointer to be sure the correct angle is still indicated.

      For perpendicular or 90° cuts, the pointer must indicate 0 degrees of angle on the arc.
3. Operating the jig saw:
   a. Secure the material to be cut using a bench vise or clamp it to a workbench or sawhorses leaving both hands free to operate the saw.
   b. Check for blade clearance; the line of cut must be free from obstructions above and below the work.
   c. Be sure the proper blade for the job is locked securely in blade holder, teeth forward and pointed up.
   d. Be certain switch is in the "off" position before connecting to the energy supply.
   e. Hold the saw handle in the one hand and guide the saw with the other hand using the guide knob (if applicable) or place hand on the upper front of the saw on the grip pad.
   f. If the saw has different operating speeds, determine the proper speed before beginning the cut.
   g. To start an outside cut, place the toe of the saw base on the edge of the material, start the motor, and move the blade into the work. Push forward and downward and guide the saw.
   h. Use constant, firm pressure on the saw to maintain a uniform forward movement. Do not overload the saw because it will result in an overheated motor or overheated blade, or cause the blade to break.
   i. Do not attempt to cut sharp curves that will twist the blade. Use narrow-bodied blades for curves and wide-bodied blades for straight cuts; use relief cuts if necessary.
   j. When making pocket or internal cuts, drill a starting hole in the waste side of the material to begin the cut.
   k. For long cuts, the switch may be locked in the "on" position by turning the saw on and depressing the switch lock with your thumb. To release the lock, pull up on the off-on switch. (This applies only to saws with a switch lock.)
   l. To prevent binding or breaking the blade, support the cut-off material until the cut has been completed.
   m. When the cut has been completed, turn off the motor and set the saw down after the motor has stopped completely.
   n. If the saw is to be removed from the cut prior to reaching the edge of the work, turn off the motor and wait until it has completely stopped before removing the saw from the cut.
   o. When finished using the jig saw, disconnect the saw from the energy supply and return it to the proper location. Depending on the storage method, it may be recommended to remove the saw blade before storing the saw.

GENERAL SAFETY PRACTICES

1. Wear approved eye protection, hearing protection, and proper clothing. Tie up loose hair and remove loose jewelry.
2. Do not operate the machine without the instructor's permission, or without instructor supervision.
3. Clamp the work securely to prevent movement or excessive vibration.
4. Work in a well-lit area.
5. Check blade for proper type, size, sharpness, and tightness in blade holder. Do not use damaged blades such as dull, bent, or cracked blades.
6. After making certain the switch is in the "off" position, connect the energy supply. For corded models the jig saw must be properly grounded to prevent injury to user.
7. Maintain a well-balanced position on both feet. Do not shift position of feet while sawing.
8. Grip handle firmly with the one hand and control turning movements with the other hand on the guide knob/upper front of the saw.
9. Place toe of base firmly on work before turning on motor.
10. Do not set saw down or remove blade from an unfinished cut until the motor has stopped.
11. When finishing a cut, support the cut-off section so it doesn't bind the blade. Bound blades may break, throwing metal pieces some distance.
12. Always disconnect the saw from energy supply when inspecting parts, making adjustments, and removing or installing blades.

COMPLETION QUESTIONS

1. The vertical adjustment locking screw is used to adjust the amount of ______________________________ exposed.
2. An angle of ______________________________ degrees should be indicated on the arc for cuts perpendicular to the surface of the material.
3. One hand should be used to grasp the ______________________________ to guide the saw.
4. The types of blades are ____________________________, ____________________________, and ______________________________.
5. The amount of blade exposed below the base of the saw should be sufficient to cut through the material at the ______________________________ point of the saw blade stroke.
6. The blade screw/quick release lever holds the ____________________________ in place on the saw.

7. One hand should be placed on the ______________________ ______________________ and used to exert a constant forward and downward pressure on the saw.

8. When the blade is properly installed, the teeth must point ___________________________ and ____________________________.

9. When starting an outside cut, the ____________________________ of the saw should be resting on the material before turning on the motor.

10. A type of internal cut called the ______________________ cut is done by ______________________ a hole to provide an opening for the blade in the waste material to begin the cut.

MULTIMETER

PART IDENTIFICATION

Identify the circled parts on the multimeter illustrated below.

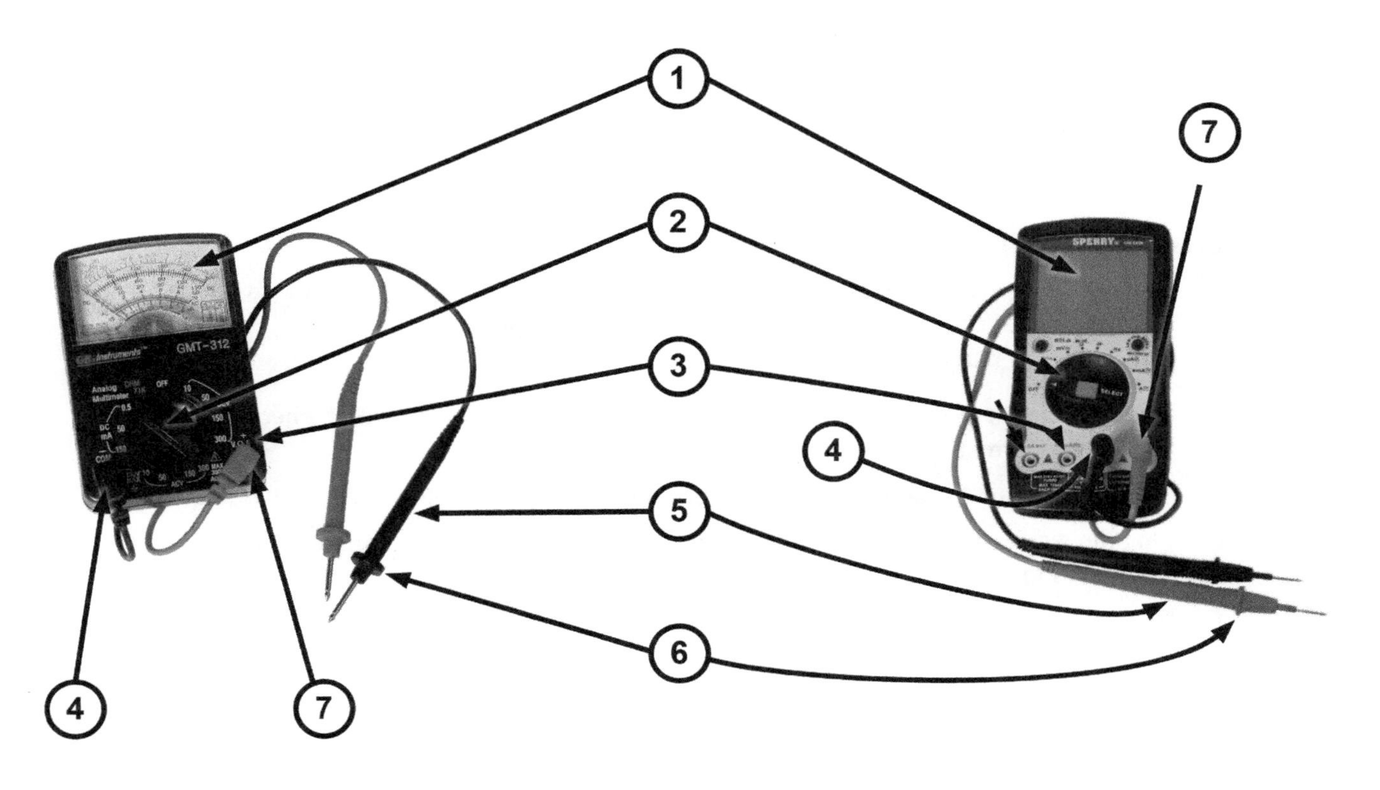

1. ______________________

2. ______________________

3. ______________________

4. ______________________

5. ______________________

6. ______________________

7. ______________________

SAFE OPERATIONAL PROCEDURES

1. Measuring Voltage
   a. Determine if you are testing AC or DC voltage.
   b. Make sure your leads are connected to the correct jacks on the multimeter: Black to common and Red to V Ω ·)))
   c. Select the DC or AC volts function by turning the function switch to DC or AC volts.
   d. Connect the common (black) test lead to the negative side of the circuit before the live (red) test lead is connected to the positive side of the circuit.

e. Read the voltage on the display.
f. Remove leads from the circuit in reverse order that they were placed.

2. Measuring Resistance
   a. Turn the power to the circuit off.
   b. Make sure your leads are connected to the correct jacks on the multimeter: Black to common and Red to V Ω ·)))
   c. Set the function range switch to Ω to test resistance.
   d. Connect the common (black) test lead before the live (red) test lead.
   e. Read the resistance on the display.
   f. Remove leads from the circuit in reverse order that they were placed.
3. Measuring Current
   a. Before testing always know what reading to expect based on the manufacture's specifications. Testing blindly is dangerous and counterproductive.
   b. Turn the power to the circuit off.
   c. Open the circuit by disconnecting or unsoldering a connection at a point where you wish to measure current.
   d. Select the DC or AC amps function by turning the function switch to DC or AC amps.
   e. Plug the test leads into the appropriate jacks, black lead into the Common jack and the red lead in to the A or mA jack.
   f. Note that the jacks used will not be the same ones used to measure voltage.
   g. Connect the tips of the probes across the break in the circuit so that the current to be measured flows through the meter. Note that this is a series connection. Never connect in parallel with the source or load, as this will cause a short circuit and damage the meter and possibly a dangerous arc flash.
   h. Turn the circuit power back on.
   i. Read the amperage on the display.
   j. Switch the circuit power off again.
   k. Remove multimeter leads from the circuit in reverse order that they were placed.
   l. If testing is finished at this point, restore the circuit by reclosing the connection.
4. Measuring Continuity
   a. Turn the power to the circuit off.
   b. Make sure your leads are connected to the correct jacks on the multimeter: Black to common and Red to V Ω ·)))
   c. Set the function range switch to ·))) to test continuity.
   d. Connect the common (black) test lead before the live (red) test lead.
   e. If there is a solid connection through the wire/component an audible tone will sound.
   f. Remove leads from the circuit in reverse order that they were placed.

GENERAL SAFETY PRACTICES

1. Wear approved eye protection, hearing protection, and proper clothing. Tie up loose hair and remove loose jewelry.
2. Do not operate the machine without the instructor's permission, or without instructor supervision.
3. Be sure to use correct personal protective equipment (PPE) in each and every situation. This means on your body (gloves, headwear) and off your body (rubber mats).
4. Always use proper lockout/tagout procedures.
5. Never work alone. Stay safe and make sure you and your partner are aware of your environment as well.
6. Before using the meter, always visually inspect it first, check the meter, test probes, and accessories for physical damage. Never use a damaged meter or test probes.
7. Always use proper jacks, switch position, and range for measurements.
8. Only use test leads that have the same voltage, category, and amperage ratings as the meter
9. Check the test leads for continuity before use.
10. Verify the meter's operation by measuring a known voltage source. Do not use the meter if it operates incorrectly.
11. Do not use the meter around explosive gas, vapor or in a damp or wet environment.
12. Disconnect circuit power and discharge all high-voltage capacitors before using meter on the circuit.
13. When using the test leads or probes, keep fingers behind the finger guards.
14. Make sure test leads do not touch while performing a measurement
15. Connect the common (black) test lead before the live (red) test lead and remove leads from the circuit in reverse order.
16. Do not apply more than the rated voltage to the multimeter, as marked on the meter.

COMPLETION QUESTIONS

1. Always use proper ________________ procedures.
2. Verify the meter's operation by measuring a ________________ voltage source. Do not use the meter if it operates incorrectly.
3. Connect the ________________ test lead before the ________________ test lead.
4. Check the test leads for ________________ before use.
5. ________________ circuit power and ________________ all high-voltage capacitors before using meter on the circuit.

6. Before using the meter, always visually inspect it first, check the meter, test probes, and accessories for physical damage. ________________ use a damaged meter or test probes.
7. Never work ________________. Stay safe and make sure you and your partner are aware of your environment as well.
8. When using the test leads or probes, keep fingers behind the ________________ guards.
9. Always use proper terminals, switch position, and range for ________________.
10. Do not use the meter around ________________ gas, vapor or in a damp or wet environment.

NAILER/STAPLER

PART IDENTIFICATION

Identify the numbered parts of the nailer/stapler illustrated below.

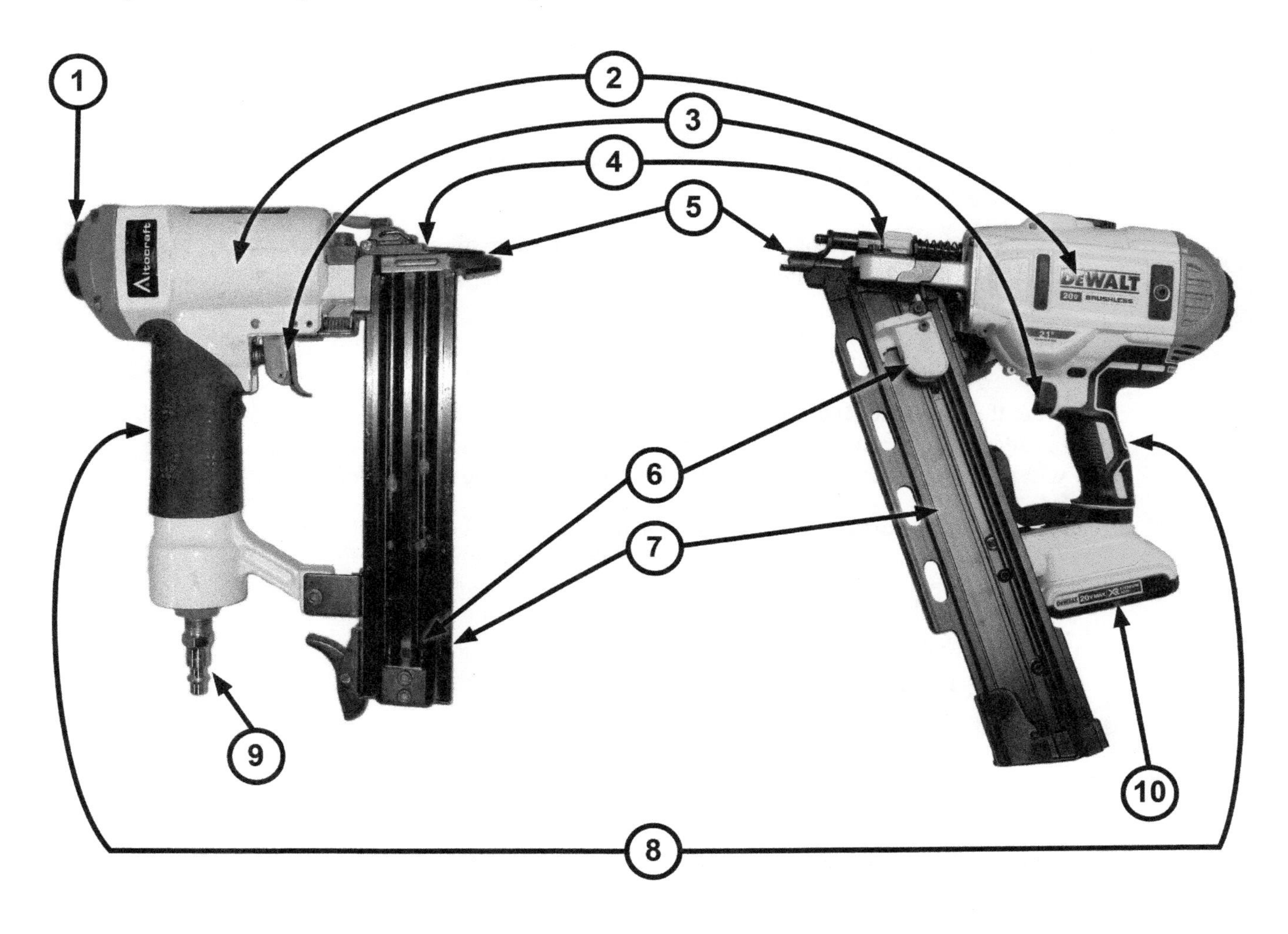

1. ______________________________

2. ______________________________

3. ______________________________

4. ______________________________

5. ______________________________

6. ______________________________

7. ______________________________

8. ______________________________

9. ______________________________

10. ______________________________

SAFE OPERATIONAL PROCEDURES

1. Study the operation, maintenance, and safety manual(s) for specific model and type of nailer or stapler.
2. Check the air pressure gauge, lines, and all connections for leaks and proper operation. Do not use tool if leaking air.
3. Operating air pressure should be between 70 and 120 pounds per square inch or PSI, 90 PSI will produce best performance for pneumatic nailers and staplers. Never exceed 120 PSI for operating a nailer or stapler.
4. Check the air inlet plug (male connector) and air coupler (female connector) to make sure the plug and coupler are of the same type and size.
5. Connect only male connectors to the tool so high-pressure air can be vented from tool when the line is disconnected. A female quick coupling could trap air in the tool leaving it live for one extra, unexpected shot.
6. Make all adjustments on the tool before loading fasteners.
7. Lubricate the tool according to manufacturer's recommendation, typically 1-2 drops per day. If oil is added into air hose with a lubricator or if you have an oil-less tool, lubricating the tool will not be necessary; however, the lubricator should be checked on a daily use basis.
8. Select size and type of fasteners for the tool and the task to be completed. Some fasteners may fit into the pneumatic tool but will not work because they are:
    a. Slightly different diameter
    b. Different magazine angle
    c. Different crown size (stapler)
    d. Different fastener spacing
9. Follow correct procedures for loading the fasteners into the magazine. Be sure that energy supply is disconnected before loading.
10. The nailer or stapler should never be used as a hammer or dropped; the tool housing may be cracked or weakened, making it unsafe. Do not engrave or stamp the main housing; this could weaken the housing unit.
11. After the tool is cleaned, lubricated, adjusted, and loaded with the proper fasteners, properly connect the tool to the energy supply for operation.
12. Carry the tool by the handle only, not by the air hose or electrical cord, and keep fingers away from the trigger. Always keep work contact element aimed toward the ground or in a safe direction and never pointed towards anyone.
13. The tool should be operated only when in contact with the material. Use caution when nailing thin materials or near corners or edges to avoid driving the fastener through or away from the material. The nailer or stapler should always be placed squarely on the surface or material to be fastened to avoid the danger of fasteners ricocheting off the surface.
14. When not using the tool or if leaving the work area, disconnect it from the energy supply.

15. Disconnect the energy supply when attempting to clear a jam or when repairing the tool.
16. Do not attempt to adjust or remove the work contacting element. If it is not working correctly, do not use the tool. This is a "safety" on most nailers and staplers and it will not fire until the element contacts the material.
17. Make sure that when you are using the nailer/stapler, the free hand that is holding the project stays out of the way. Keep fingers at least 4" away from the circumference area to be fastened as the fasteners could shoot out at odd angles if you are not holding it straight.
18. When work is completed, disconnect the unit from the energy supply. If it was a pneumatic tool, also shut off the air supply to the air hose, and vent the compressed air from air hose with an air nozzle.
19. Remove the fasteners from the tool, clean tool, and place in proper storage.

GENERAL SAFETY PRACTICES

1. Wear approved eye protection, hearing protection, and proper clothing. Tie up loose hair and remove loose jewelry.
2. Do not operate the machine without the instructor's permission, or without instructor supervision.
3. Always assume the tool is loaded (contains fasteners).
4. Never point the tool at anyone, even if the energy supply is disconnected.
5. Disconnect energy supply when not in use or when leaving the work area.
6. Carry the tool only by the handle, not by the air hose or cord, and never place your finger on the trigger.
7. Make sure the work contacting element is in good working order and in contact with the material before depressing the trigger.
8. Use matching air connectors and couplings.
9. Select correct fasteners manufactured for the tool.
10. Do not remove or tamper with the work contacting element.
11. Use only regulated air pressure; never use bottled air or gases to power the tool.
12. Use only recommended air pressure; under-pressure operation may be as dangerous as over-pressure operation.
13. The nailer or stapler is a tool, not a toy.

COMPLETION QUESTIONS

1. The recommended air pressure for most air nailers and staplers is ________________ PSI.
2. Always assume the tool is __________________________.

3. The energy supply for the pneumatic nailer or stapler should be regulated ____________________ ____________________, not bottled air or gas.

4. Carry the tool only by the ____________________ with the work contacting element pointed in a safe direction.

5. The __________ ____________ ______________ should always be in contact with the material before the tool is fired.

6. Only ____________________ connectors should be fitted to the tool.

7. Stamping or engraving on the ____________ ____________ could weaken the tool, causing it to be unsafe.

8. The fasteners are held in the ____________________ portion of the tool.

9. The tool should be ____________________ daily unless oil is added directly into the air hose.

10. Another name for an air tool is a ____________________ tool.

OSCILLATING MULTI-TOOL

PART IDENTIFICATION

Identify the circled parts on the oscillating multi-tool illustrated below.

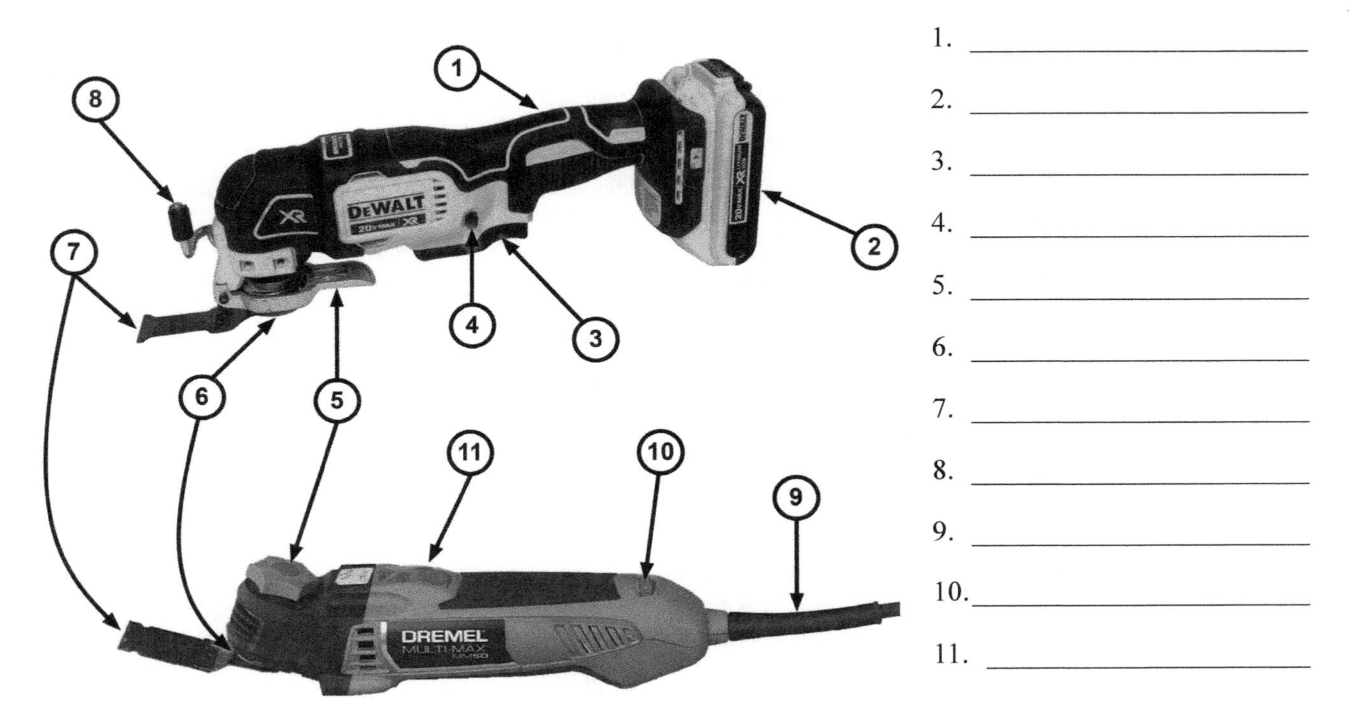

SAFE OPERATIONAL PROCEDURES

1. This tool works by oscillating rapidly. You should not force or overexert the tool. Allow the tool to cut or do its job in order to get the best possible performance.
2. Secure small workpiece to the workbench.
3. Retract the power switch to its off position. Connect the tool into the energy source
4. Turn on the tool and allow it to reach operating speed.
5. Apply gentle forward pressure to allow the tool to begin cutting.
6. When finished, turn off the tool and allow it to completely stop before setting it down.
7. Disconnect the tool from the energy source.
8. Prior to storing the tool, remove the attachment and wipe it clean. Store the tool in a clean, dry, safe location out of reach of children and other unauthorized persons.

GENERAL SAFETY PRACTICES

1. Wear approved eye protection, hearing protection, and proper clothing. Tie up loose hair and remove loose jewelry.
2. Do not operate the machine without the instructor's permission, or without instructor supervision.
3. Protect your lungs. Wear a face or dust mask during operation.
4. Operations such as wire brushing, polishing or carving are not recommended to be performed with this power tool.
5. Do not use accessories which are not specifically designed and recommended by the tool manufacturer.
6. The rated speed of the accessories must be at least equal to the maximum speed marked on the power tool.
7. Do not use a damaged tool.
8. Hold power tool by insulated gripping surfaces only.
9. Use clamps to support workpiece whenever practical. Never hold a small workpiece in one hand and the tool in the other hand while in use.
10. Never lay the power tool down until the accessory has come to a complete stop.
11. Regularly clean the power tool's air vents.
12. Do not operate the power tool near flammable materials.
13. When tool is pinched, snagged or when interrupting a cut for any reason, switch off the power tool and hold the power tool motionless until the tool comes to a complete stop.
14. Use extra caution when making a "plunge cut" into existing walls or other blind areas.
15. Keep hands and fingers safely away from the oscillating angled scraper, blades, sanding pad and sandpaper when operating the tool.

COMPLETION QUESTIONS

1. This tool works by ________________ rapidly.
2. Secure ________________ workpiece to the workbench.
3. Retract the ________________ switch to its off position.
4. Hold power tool by ________________ gripping surfaces only.
5. Never lay the power tool down until the ________________ has come to a complete stop.
6. Do not operate the power tool near ________________ materials.
7. Apply gentle ________________ pressure to allow the tool to begin cutting.

8. Operations such as wire brushing, polishing or carving are ________________ recommended to be performed with this power tool.

9. Keep ________________ ________________ ________________ safely away from the oscillating angled scraper, blades, sanding pad and sandpaper when operating the tool.

10. Do not use ________________ which are not specifically designed and recommended by the tool manufacturer.

OXY-FUEL TORCH

PART IDENTIFICATION

Identify the numbered parts of the oxy-fuel torch illustrated below.

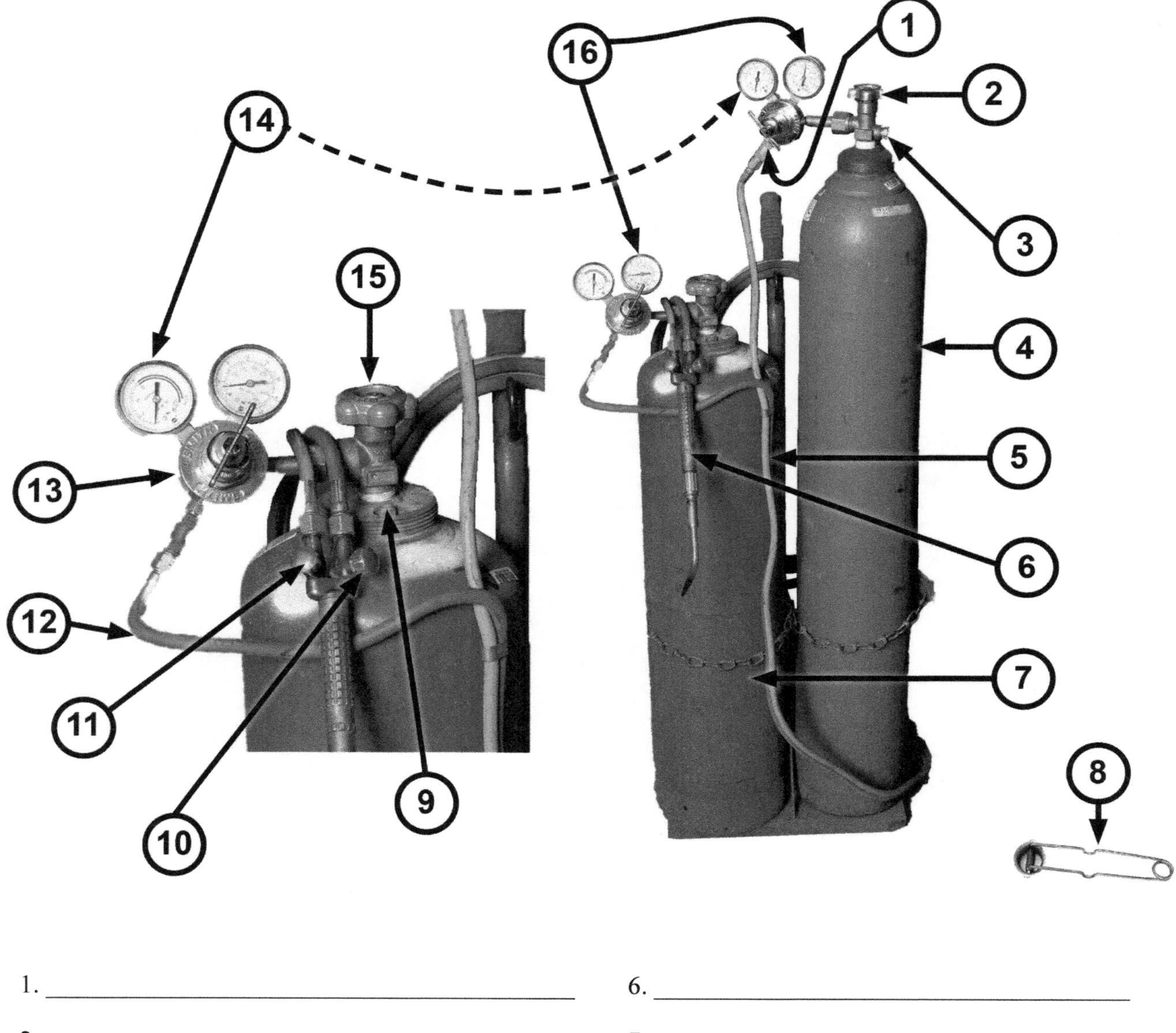

1. ______________________

2. ______________________

3. ______________________

4. ______________________

5. ______________________

6. ______________________

7. ______________________

8. ______________________

9. ______________________

10. ______________________

11. ______________________________ 14. ______________________________

12. ______________________________ 15. ______________________________

13. ______________________________ 16. ______________________________

SAFE OPERATIONAL PROCEDURES

1. Setting up equipment:
    a. Fasten oxygen and fuel gas cylinders securely in an upright position. They should be chained to the wall or chained to a cart.
    b. "Crack" the cylinder valves to blow out dust.
    c. Attach the regulators to the cylinder valves. Fuel gas regulator nuts have left-hand threads.
    d. Be sure the regulator nuts fit the cylinder valves properly. Do not force the threads.
    e. Connect the end of the oxygen hose (green) and the fuel gas hose (red) to the regulator and the torch body. Torch connection are normally marked "oxy" and "acet". It is impossible to attach hoses incorrectly since oxygen connections have right hand thread and fuel gas connections have left hand threads.
    f. Select the cutting tip or blow pipe that is suitable for the job you are to do. Refer to the manufacturer's recommendation for specific sizes for given applications.
    g. Hand tighten the welding tip or cutting torch on the torch body; make sure that all sealing rings are in place.
    h. Be sure the regulator valves are off. Turn the adjusting screws counter clockwise until they are loose in the threads.
    i. Open the oxygen cylinder valve very slowly until the gauge reaches its maximum reading; then turn the cylinder valve all the way open.
    j. Open the fuel gas cylinder valve slowly, only one-half turn in case of emergency and a quick shutdown is needed.
    k. Test connections suspected of leaking with non-oil-based soap suds.
2. Lighting the torch:
    a. Open fuel gas cylinder valve one-half turn and leave the wrench in position at all times if your tank uses a wrench. This will ensure for a quick shut-off in case of an emergency.
    b. Open the oxygen cylinder valve all the way open.
    c. Open the regulator valves by turning the adjusting screw clockwise until the proper pressure is obtained for the job.
    d. Open the fuel gas valve at the torch 1/8 turn.
    e. Use a friction lighter and light the torch. NEVER USE A MATCH OR FLAMMABLE LIGHTER.
    f. Adjust the fuel gas flow, by turning the torch fuel gas valve, until the flame just starts to produce black smoke around its edges; then increase fuel gas flow just enough to get rid

of the black smoke. Another method is to open the fuel gas valve until the flame leaves the end of the tip and then reduce the fuel gas flow until the flame comes back to the tip.

g. Place the appropriate shaded goggles or face shield over the eyes.

h. Open oxygen torch valve until a neutral flame is produced. A neutral flame has no feather and has a blunt or rounded inner cone.

3. Cutting with oxy-fuel:

   a. Select the proper hose pressure for the job. Common pressures for cutting range from 30 to 40 PSI for oxygen and 3 to 5 PSI for acetylene (common fuel gas).

   b. Open the blowpipe fuel gas valve about 1/8 turn; then light the torch with a friction lighter.

   c. Increase the fuel gas until the flame is just ready to leave the tip.

   d. Open the oxygen preheat valve until a neutral flame is obtained.

   e. Hold the torch with both hands with the right hand near the lever.

   f. Keep the inner cones of the neutral preheat flames about 1/16" above the metal.

   g. Allow a molten puddle to form on the edge; then press the oxygen lever.

   h. Tilt the torch so the flame leads slightly and move along as fast as possible from right to left.

4. Turning off the torch:

   a. First close the fuel gas torch valve, then close the oxygen valve. Closing the fuel gas valve first reduces the chance of allowing unburned fuel gas to escape and be ignited accidentally.

   b. First close both cylinder valves and then open the fuel gas and oxygen valves on the torch body, one at a time. Allow the gas in each line to escape and see the line pressures go to zero on the regulator gauge.

   c. Close the fuel gas and oxygen torch valves after the gas in each line has been released. This is to prevent the reverse flow of gas into an opposite line.

   d. Release the line pressure-adjusting screws on the fuel gas and oxygen regulators.

GENERAL SAFETY PRACTICES

1. Wear approved eye protection, hearing protection, and proper clothing. Tie up loose hair and remove loose jewelry.

2. Do not operate the machine without the instructor's permission, or without instructor supervision.

3. Always wear safety glasses with a MINIMUM shade 5 face shield or shaded goggles when welding or cutting.

4. Never cut or weld a container until it has been made safe. Welding or cutting on a container when the contents is unknown could result in an explosion.

5. Wear gloves and clean coveralls when welding or cutting.

6. Turn down pants cuffs and wear high-top shoes when cutting or welding.
7. Do not cut galvanized metal indoors. A toxic gas is given off.
8. Never work with oxy-fuel equipment that may be defective.
9. Never use oil or grease around the oxygen cylinder, regulator, or torch.
10. Inspect hoses and test the connections for leaks using only non-oil-based soap suds.
11. Never repair a hose with tape.
12. Never light the torch with a match or lighter.
13. Keep hot metal, the flame, sparks, hot slag, and sharp edges away from the hoses.
14. Keep the protector caps on cylinders not in use. Never drop cylinders.
15. Never use acetylene pressures above 15 PSI.
16. Never open fuel gas cylinder valves more than 1/2 turn. This will allow the cylinder to be shut off quickly in the event of a fire.
17. Never release oxygen or fuel gas in a confined area. The fuel gas may ignite and the concentrated oxygen may cause clothing and other combustible material to burn very quickly.
18. Never point the torch toward anyone and never set an open-flamed torch down. Always shut it down first.
19. Never use oxygen or fuel gas to blow dirt off clothing. Clothing saturated with oxygen or fuel gas will burn very quickly.
20. A flashback arrester is recommended. If a flashback should occur, turn off the oxygen valve on the torch and close the fuel gas cylinder valve immediately. Cool the torch and determine the cause of the flashback.
21. Clear the area of all combustible materials before lighting the torch.
22. Do not tamper with fusible plugs or safety valves on cylinders. They are safety valves that open in case of excessive internal tank pressures.
23. Never use boiling water to thaw ice from the outlet of a fuel gas cylinder. Boiling water can melt the fusible plugs.
24. Never leave hot metal where others may be burned by it.
25. Turn on the ventilation fan before lighting the oxy-fuel gas torch.

COMPLETION QUESTIONS:

1. Oxygen and fuel gas cylinders should be securely fastened in a ______________________________ position.
2. ______________________________ the cylinder valves to blow out the dust before attaching regulators.

3. The fuel gas regulator nut has ______________________ hand threads.
4. In turning off the regulators, the adjusting screws must be turned ______________________.
5. The fuel gas cylinder valve should be opened only ______________________ turn(s).
6. Use __________________________ __________________________ to test connections for leaks.
7. A neutral flame has no ______________________ and has a ______________________ inner cone.
8. Before lighting the torch, the blowpipe fuel gas valve should be opened about __________________________ turn.
9. The __________________________ __________________________ valve should be shut off first to turn off the torch.
10. __________________________ is the most common fuel gas.

PLASMA CUTTER

PART IDENTIFICATION

Identify the circled parts on the plasma cutter illustrated below.

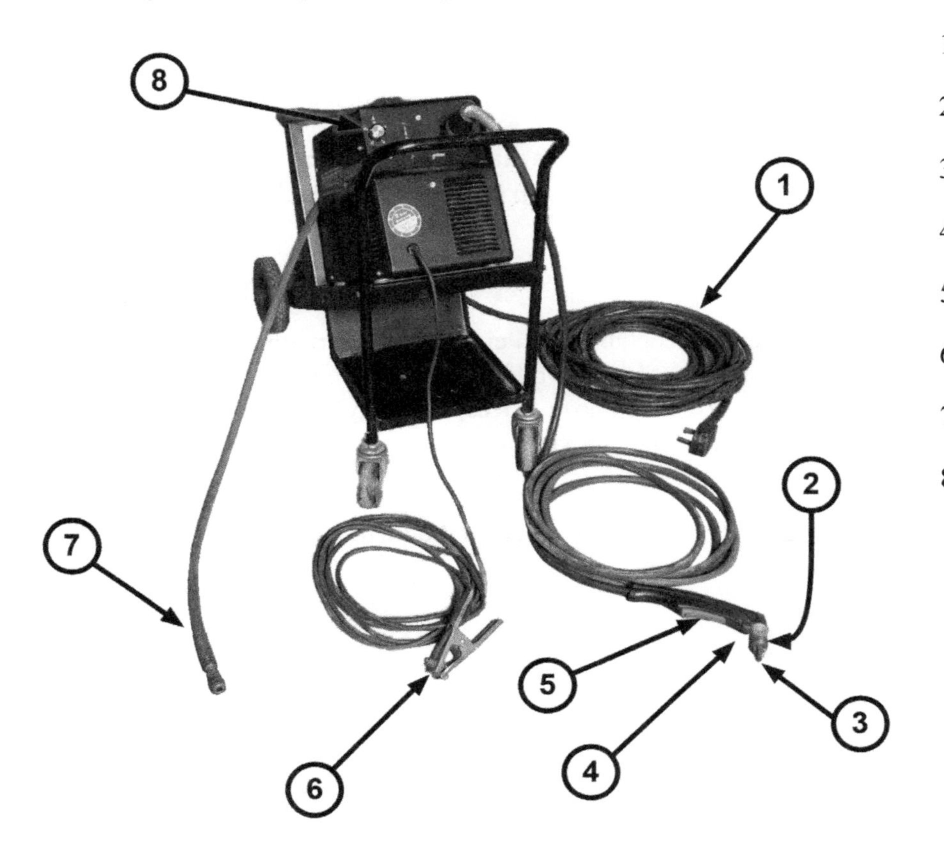

1. ____________________
2. ____________________
3. ____________________
4. ____________________
5. ____________________
6. ____________________
7. ____________________
8. ____________________

SAFE OPERATIONAL PROCEDURES

1. To activate the plasma cutter, make sure the air pressure is around 70 PSI for most plasma cutter units and the ground clamp is attached to the work piece.
2. Turn the plasma cutter on and adjust the amperage to the manufacturer's specifications for the thickness of metal to be cut.
3. Position the shielding cup over the metal, press the igniter button and allow the arc to become established. Next, move the arc over the cut line and make the cut.
4. The shielding cup and constricting nozzle should be held approximately 1/8" to 1/4" above the metal being cut. The operator should avoid dragging the constricting nozzle and shielding cup on the metal when making the cut unless they are specifically designed to touch the base metal while cutting.

5. The thicker the metal, the slower the travel speed must be to get a good cut and vice versa. The quality of the cut usually decreases as the metal thickness increases and the travel speed decreases.
6. A guide bar may be used to help achieve good straight cuts.
7. Always make cuts on the waste side of the cut line.
8. Cuts with the plasma cutter may be made by moving forward, backward, or sideways.
9. Determine which direction is easiest for you and use that procedure as often as possible.
10. Always move the plasma cutter as fast as possible when making a cut. This increases time efficiency, improves the cut quality, and reduces the buildup of dross on the back side of the cut.
11. Always turn the plasma cutter off before laying the torch down and leaving the work area.
12. If the quality of the cut deteriorates to an unacceptable level either the constricting nozzle, the electrode, or both may need to be changed. The electrode on most plasma arc cutters will have a life of about twice the life of the constriction nozzle. Keep a supply of constricting nozzles and electrodes on hand as they deteriorate quickly during continuous use. Turn the plasma arc cutter off to put on replacement parts.
13. Keep the plasma arc cutter torch leads and ground lead stored so they will not be cut or damaged when not in use.

GENERAL SAFETY PRACTICES

1. Wear approved eye protection, hearing protection, and proper clothing. Tie up loose hair and remove loose jewelry.
2. Do not operate the machine without the instructor's permission, or without instructor supervision.
3. Use an approved helmet with minimum #5 shaded lens. The shaded lens needed to adequately protect the eyes and varies by the thickness of the metal being cut and the intensity of the arc required to make the cut. Follow the manufacturers recommendation for selecting an appropriate shaded lens for given plasma arc cut.
4. Hearing protection should be worn when operating the plasma arc cutter.
5. Always wear protective clothing suitable for plasma cutting. Wool or cotton clothing, close-toed shoes (high top leather shoes recommended), leather gauntlet gloves, welding coat or leathers. DO NOT wear clothing made of synthetic fibers when plasma cutting. Some synthetic fibers are highly flammable. All this must be worn to prevent burns from ultraviolet and infrared rays emitted while arc welding.
6. Guard against the use of damp or wet clothing when plasma cutting. The use of such clothing increases the possibility of electrical shock.
7. Never plasma cut while standing in water or on damp ground. Dampness on the body increases the chance of electrical shock when plasma cutting.

8. Do not talk with observers while operating machines. Give the plasma cutter your full attention.

9. Do not carry matches, butane or propane lighters or other flammables in your pockets while plasma cutting.

10. Make sure that work area is well ventilated when using the plasma arc cutter. It will generate a lot of fumes and therefore must be well ventilated. Position yourself so there will be minimum exposure of fumes during the cutting process.

11. Fumes from some metals can be toxic and, in some cases, fatal. Avoid cutting these metals with the plasma arc cutter in the school setting. When encountered in the industrial setting, special care must be taken to avoid breathing fumes given off by these metals: Antimony, Chromium, Mercury, Arsenic, Cobalt, Nickel, Barium, Copper, Selenium, Beryllium, Lead, Silver, Cadmium, Manganese, Vanadium.

12. Chlorinated solvents and cleaner vapors in the presence of plasma arc cutter generates a toxic phosgene gas. Avoid plasma arc cutting use in areas which house chlorinated solvents and cleaners.

13. Hydrogen gas may be formed and trapped when cutting aluminum in the presence of water. Trapped hydrogen gas in the presence of an arc will ignite and explode, make sure fumes are well ventilated when cutting aluminum.

14. Use a cutting table which has a down draft to capture fumes. A cutting table with water filtration is also recommended for plasma arc cutting.

15. If plasma cutting over an open barrel with a grate be aware that the fume plume will be directed back toward the operator. Avoid this condition if at all possible, otherwise limit the exposure to fumes to short durations.

16. Do not use the plasma cutter to cut on containers that have held combustible materials.

17. Never use the plasma cutter in areas where combustible or explosive gases or materials are located.

18. Shield others from the light rays and fumes produced by the plasma cutter. Keep the welding curtain in place at all times to protect others from arc flash.

19. Keep the plasma cutting area clean and free of tools, scrap metal, and water.

20. Make sure the work area is free of flammable, volatile, or explosive materials. (Ex. propane, gasoline, grease, paper and coal dust).

21. Protect plasma cutting cables from sparks, hot metal, open flames, sharp edges, oil, and grease. Do not use cables with frayed, cracked or bare spots in the insulation.

22. Never plasma cut with the cables coiled over the shoulders.

23. Compressed air used in the plasma cutter should be dry or the cutter will not yield a quality cut, or it could not cut at all. An auxiliary air filter may be place in the compressed air line to condition the air for a plasma cutter.

24. Report to supervisor at once if the cable connection, cable, cable terminals at the plasma cutting machine, ground clamps, or lugs get hot.

25. Never touch any parts on the plasma arc cutter that are electrically connected. The plasma cutter uses high amperage and produces high voltage which can cause severe or fatal electrical shock.

26. Disconnect the electrical power before performing any service or repair on the plasma cutter.

27. Use tongs or pliers to handle hot metal after it has been plasma cut. Completely submerge metal in water when cooling, this prevents steam from burning you. Cool and store any hot metal before leaving the work area.

28. Avoid plasma cutting directly on concrete floors. Residual moisture in the concrete may be turned to steam resulting in the concrete exploding.

29. Use a fire blanket to smother clothing fires. Use a dry chemical type "C" extinguisher to put out an electrical fire.

COMPLETION QUESTIONS

1. Use an approved helmet with minimum _______________ shaded lens.

2. The shielding cup and constricting nozzle should be held approximately _______________ above the metal being cut.

3. A _______________ _______________ may be used to help achieve good straight cuts.

4. Always turn the plasma cutter off before laying the _______________ down and leaving the work area.

5. Never plasma cut while standing in _______________ or on damp ground.

6. Do not carry _______________, butane or propane lighters or other flammables in your pockets while plasma cutting.

7. Fumes from some metals can be _______________ and, in some cases, _______________.

8. Hydrogen gas may be formed and trapped when cutting _______________ in the presence of water.

9. If plasma cutting over an open barrel with a grate be aware that the _______________ _______________ will be directed back toward the operator.

10. Compressed air used in the plasma cutter should be _______________ or the cutter will not yield a quality cut, or it could not cut at all.

PORTABLE BELT SANDER

PART IDENTIFICATION

Identify the numbered parts of the portable belt sander illustrated below.

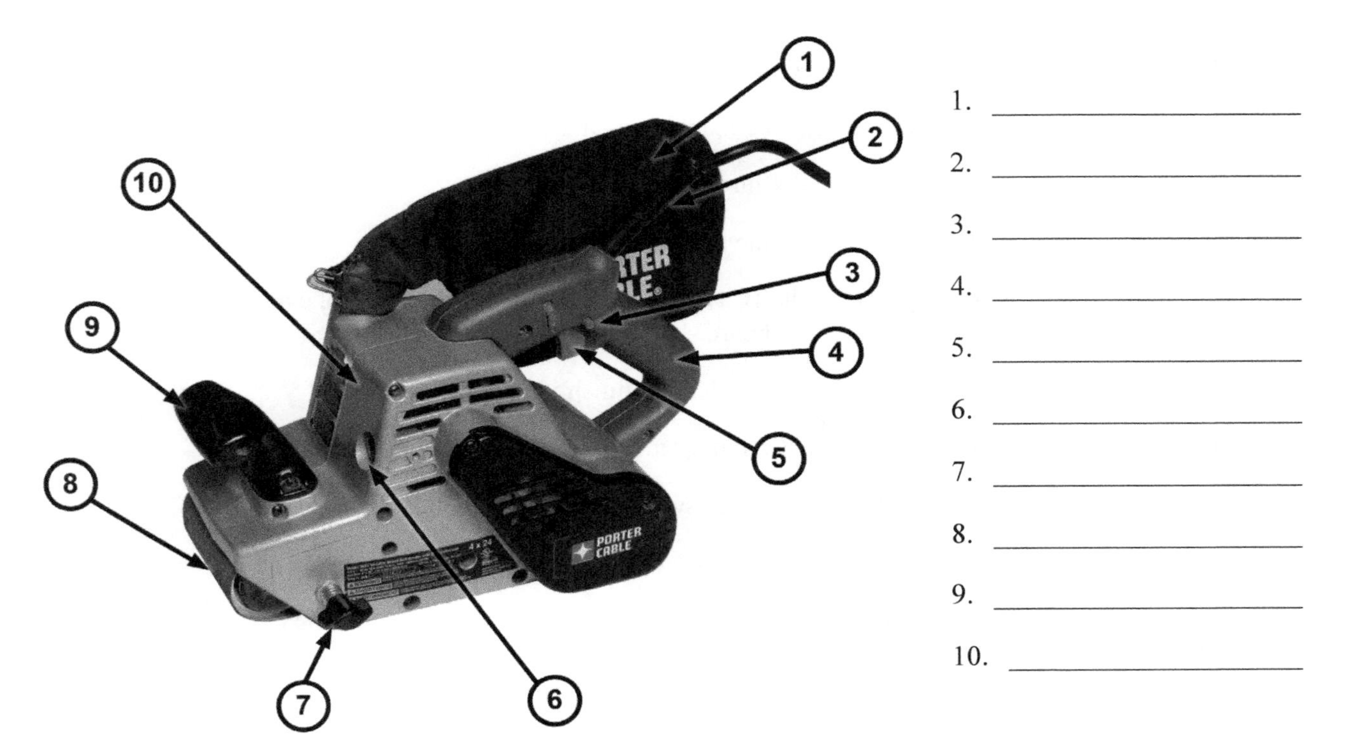

SAFE OPERATIONAL PROCEDURES

1. Installing the belt:
   a. Make sure power source is disconnected before removing or replacing the belt.
   b. Check condition of belt to see that it is not frayed on the edges, torn, or filled with glue and sanding dust.
   c. The belt turns on two pulleys or spindles, one of which is adjustable for tightening and aligning the belt.
   d. Select the correct size belt for the specific machine and coarseness of grit according to the job to be sanded.
   e. Loosen or release belt tightening mechanism.
   f. Place belt on pulleys; note arrow on inside of sanding belt for proper direction of travel.
   g. Tighten belt so it has tension, keeping it snug against the pulleys and sander base.
   h. Check the off-on switch to see that it is "off" and not locked in the "on" position.
   i. Connect the sander to energy source.

j. With sander upside-down and well supported, pull trigger switch and check belt alignment or tracking. Turn tracking adjustment knob to make sure the belt is tracking on the center of the front pulley. If off center, the belt may cut through the sander housing

2. Before starting sanding operation, check the dust bag and empty if it contains dust.
3. Clamp material to be sanded in vise or to table.
4. Hold the sander in both hands with the one hand on the switch or back handle and the other hand on the front handle. Start sander above the work.
5. Allow rear of belt to touch first, leveling machine as it is moved forward.
6. Using the weight of the sander, sand with the grain.
7. Raise the sander slightly at the end of the stroke.
8. Begin the second stroke by lapping halfway across the width of the first stroke and continuing this procedure across the width of the board.
9. Use progressively finer grit belts until the desired finish is obtained.
10. Do not pause in any one spot. Keep the sander moving over the material at all times.
11. Lift the sander off the work before allowing switch to go to the "off" position.
12. Wait until the motor is completely stopped before placing the sander on its side on the bench.
13. When the work is completed, disconnect the sander from the power source, clean the machine, and empty the dust bag before storing it.

GENERAL SAFETY PRACTICES

1. Wear approved eye protection, hearing protection, and proper clothing. Tie up loose hair and remove loose jewelry.
2. Do not operate the machine without the instructor's permission, or without instructor supervision.
3. Check motor switch to see that it is off before connecting to energy source.
4. Do not attempt to take material off too fast by excessive pressure and overworking the motor.
5. Disconnect the energy source from the sander when not in use.
6. Keep the sander clean and well lubricated.

COMPLETION QUESTIONS

1. Portable belt sanders are sized by the ____________________________ of the belt.
2. Before connecting the energy source, make sure the switch is in the ______________________ position.

3. The belt is adjusted back and forth on the front pulley by turning the ________________________ knob.

4. In starting the sanding operation, set the ____________________________ of the sander down first.

5. Always sand in the same direction of the ____________________________, moving the sander back and forth over a wide area.

6. In finishing a job, use progressively ___________________________ ___________________________ sanding belts until the desired finish is obtained.

7. When not sanding, set the sander on its __________________________ and near the center of the bench.

8. Keep the machine ____________________________ and well ____________________________ for long life.

9. Never set the sander down until the motor is completely ____________________________.

10. Place the belt on the pulley with the arrow ____________________________ the direction of travel.

PORTABLE BUFFER

PART IDENTIFICATION

Identify the circled parts on the portable buffer illustrated below.

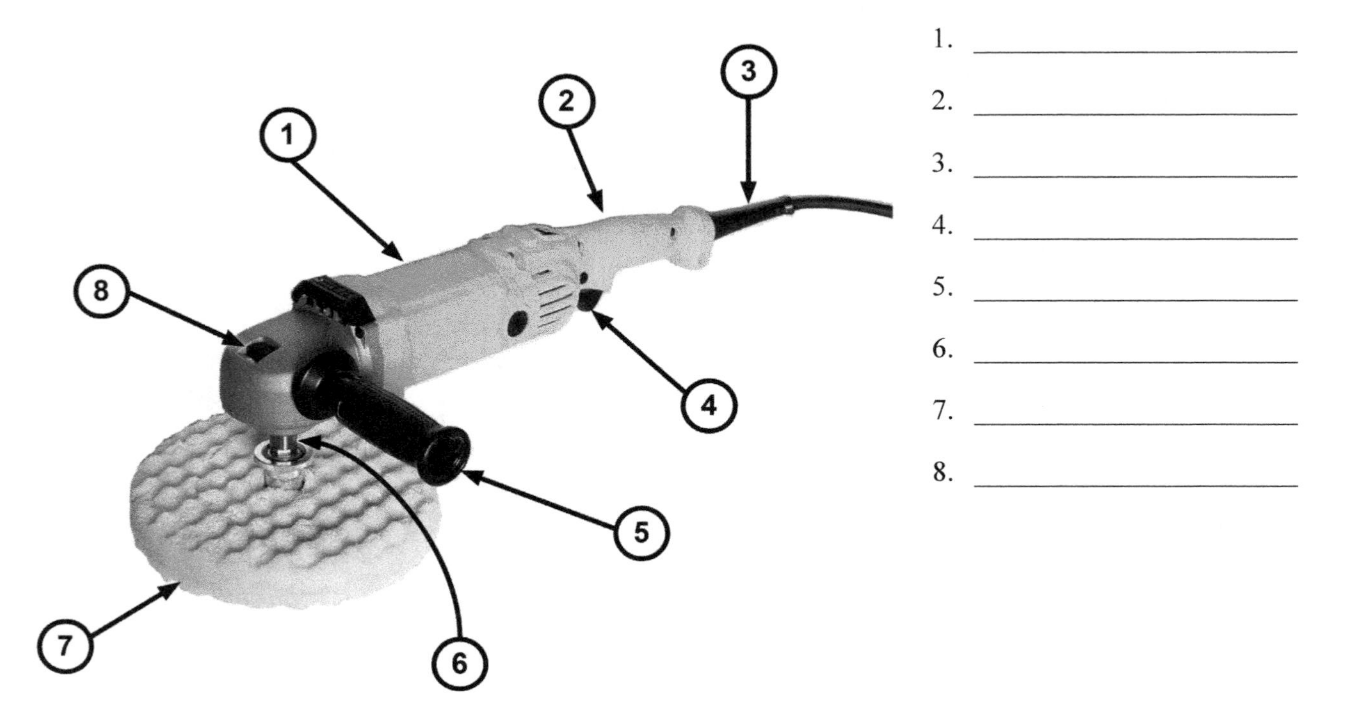

1. ______________________
2. ______________________
3. ______________________
4. ______________________
5. ______________________
6. ______________________
7. ______________________
8. ______________________

SAFE OPERATIONAL PROCEDURES

1. Before using the portable buffer, inspect for damage or disrepair.
2. Turn off the portable buffer and disconnect the power source prior to making adjustments.
3. Always clean the portable buffer work area upon completion of your work.
4. Review and understand information provided in the portable buffer operators manual.
5. Use the portable buffer in a well ventilated and well-lit area.
6. Always use two hands to operate the portable buffer.
7. Make sure the portable buffer has come to a complete stop before setting it down.

GENERAL SAFETY PRACTICES

1. Wear approved eye protection, hearing protection, and proper clothing. Tie up loose hair and remove loose jewelry.

2. Do not operate the machine without the instructor's permission, or without instructor supervision.
3. Wear a dust mask when operating the portable buffer.
4. Do not operate the portable buffer with gloves.
5. Ensure the portable buffer does not get in the path of the cord.
6. Check to be sure all workspaces and walkways are clear of slip/trip hazards.
7. Use the appropriate type of pad and polishing compound for the task.
8. Only use the face of the buffing pad, not the edge.
9. Switch off the tool when work is finished, leave the tool clean and safe for the next person.

COMPLETION QUESTIONS

1. Always use ________________ hands to operate the portable buffer.
2. Make sure the portable buffer has come to a complete ________________ before setting it down.
3. Never use the portable buffer without getting the instructors ________________.
4. Only use the ________________ of the buffing pad, not the edge.
5. Tie up ________________ hair and remove loose jewelry.
6. Ensure the portable buffer does not get in the path of the ________________.
7. Wear a dust mask when ________________ the portable buffer.
8. Turn off the portable buffer and ________________ the power source prior to making adjustments.
9. Use the portable buffer in a well ventilated and well- ________________ area.
10. Use the appropriate type of ________________ and polishing compound for the task.

PORTABLE DRILL

PART IDENTIFICATION

Identify the numbered parts of the portable drill illustrated below.

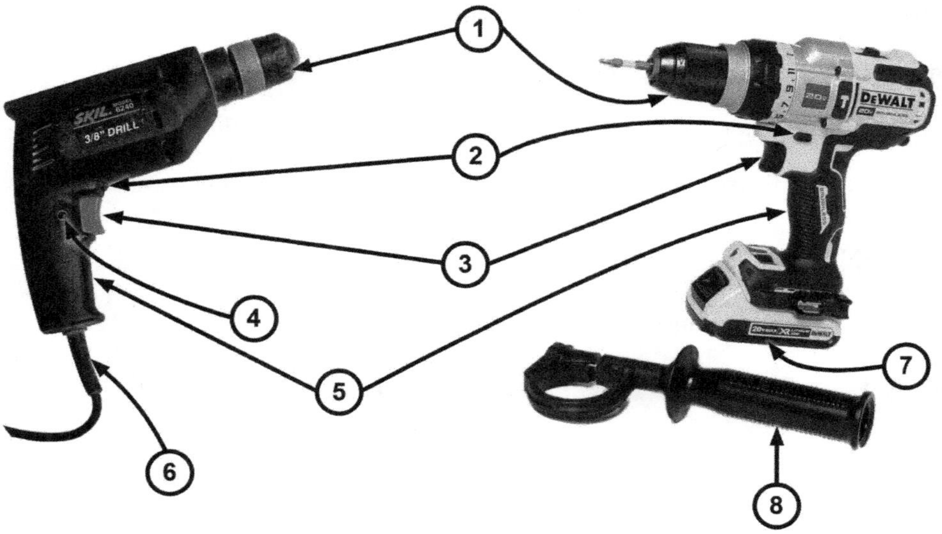

1. ____________________
2. ____________________
3. ____________________
4. ____________________
5. ____________________
6. ____________________
7. ____________________
8. ____________________

SAFE OPERATIONAL PROCEDURES

1. Select the proper bit for material to be drilled and make sure it is properly sharpened.
2. Driver bits are useful when using a portable drill to drive screws; they do not drill holes of any kind.
3. Tighten chuck securely in all three tightening locations so that the bit will run true.
4. Locate exact point of desired penetration and mark with a center punch or awl.
5. Securely fasten all materials so that they cannot twist or turn during the drilling process.
6. Small size twist drill bits should run at a faster RPM than large size twist drill bits.
7. Let drill come to full speed before starting to drill.
8. Drill with even, steady pressure and let the drill do the work.
9. When drilling deep holes, withdraw the drill several times to clear the cuttings or chips.
10. Hold drill at the correct angle while drilling.
11. As the twist drill bit starts to break through the material, ease off on the pressure to prevent grabbing or splintering.
12. Apply a suitable coolant or lubricant if necessary.

13. Clamp scrap material on the back side of material to be drilled so it will not grab or splinter when the twist drill bit breaks through.
14. Raise drill soon after it starts cutting and inspect the impression to make sure it is centered.
15. Twist drill bits are made and sharpened to operate in the right-hand direction.
16. When you are planning on drilling holes larger than ¼" with a twist drill bit, make sure to start with a smaller pilot bit and expand the hole in stages; large bits have a tendency to walk, and it might take very long to get through the material.
17. Disconnect the drill from the energy supply and remove the bit from the drill as soon as work has been completed.
18. Larger twist drill bits should be used at slower speeds.
19. Variable speed drills can operate from 0 to 1,000 RPM.

GENERAL SAFETY PRACTICES

1. Wear approved eye protection, hearing protection, and proper clothing. Tie up loose hair and remove loose jewelry.
2. Do not operate the machine without the instructor's permission, or without instructor supervision.
3. Disconnect drill from energy supply when not in use, before servicing and cleaning, and when changing bits.
4. Before connecting to energy supply, make sure the switch is in the "off" position.
5. When using a drill with a keyed chuck, make sure the key has been removed from the chuck before starting the drill.
6. Be sure the drill is properly grounded when connected to the power outlet or is double insulated.
7. Do not use a drill which has a broken or damaged part(s).
8. While using a large drill, brace the body well to prevent injury.
9. Use a brush to remove chips.
10. Keep attention focused on the work, keep good footing, and maintain good balance
11. Note position of cord to avoid drilling into it or getting it wrapped around the twist drill bit.
12. Make certain the drill will not injure someone working on the other side of the work.
13. Never operate the portable drill while standing in water or on a wet floor.

COMPLETION QUESTIONS

1. Be sure the switch is ____________________________ before connecting the drill to the energy supply.
2. A ____________________ ____________________ or ____________________________ can be used to establish the starting point for the twist drill.
3. Larger twist drill bits should be used at a ____________________________ speed.
4. A pilot hole is used so less pressure is required when drilling ____________________________ holes.
5. When drilling deep holes, withdraw the twist drill bit several times to clear the ____________________________.
6. Disconnect the drill from __________________________ __________________________ before removing or installing twist drill bits.
7. Check to see if the _________________________ has been removed from the chuck before starting the drill.
8. Use a ____________________________ to remove chips.
9. Twist drill bits are made to operate in a ____________________________ hand direction.
10. Variable speed drills can operate at speeds from ____________________________ to 1,000 RPM.

PORTABLE METAL-CUTTING BAND SAW

PART IDENTIFICATION

Identify the numbered parts of the portable metal-cutting band saw illustrated below.

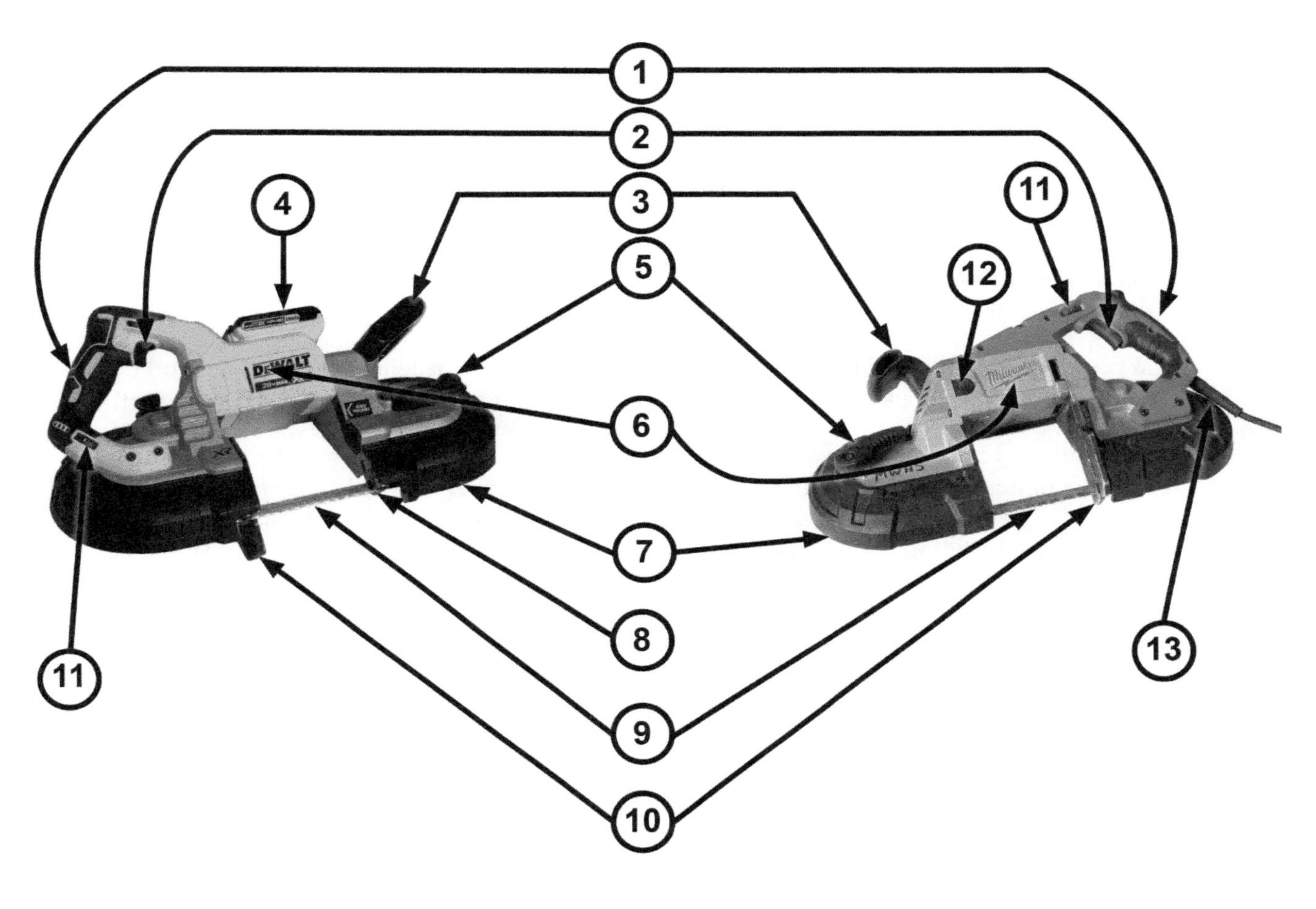

1. ______________________________

2. ______________________________

3. ______________________________

4. ______________________________

5. ______________________________

6. ______________________________

7. ______________________________

8. ______________________________

9. ______________________________

10. ______________________________

11. ______________________________

12. ______________________________

13. ______________________________

SAFE OPERATIONAL PROCEDURES

1. Study the operation, maintenance, and safety manual(s) for the specific model and type of saw.
2. Select a blade that is sharp and has 18-24 teeth per inch. Select a blade which will allow at least two teeth to be engaged in the material thickness. The thinner or the harder the material, the finer the blade teeth. The thicker or the softer the material, the coarser the blade teeth.
3. To remove the worn blade, turn the blade release handle clockwise to release the blade tension. Remove the blade first from the pulleys and then from the blade guides.
4. Before installing a blade, clean chips and wax from the blade guides and pulley tires. Insert blade in blade guides first and then position on pulleys. Place the blade in the saw with the exposed teeth pointing toward the handle or back part of the saw.
5. When the new blade is in place, turn the blade release handle counterclockwise. Check the blade to see that it is snug and fitting into the guide rollers.
6. When a worn or broken blade is replaced, do not start a new blade in a partially cut workpiece as the set of a new blade is thicker and will be damaged if allowed to enter an old saw kerf.
7. If the saw has two-speed or variable speed control, select the correct saw speed according to the type of material to be cut. Blade speed is SFPM (Surface Feet Per Minute) and will range from 80 to 240 SFPM. The general rule is the harder the material such as high-speed steel, stainless steel, chrome, or tungsten steel, the slower the blade speed (80-150 SFPM). Metals such as aluminum, brass, copper, soft bronze, and low carbon steel may be cut at high speed, 150 to 240 SFPM.
8. If the saw has an adjustable blade guide, adjust blade guide according to size of material, keeping guide as close to material as possible.
9. Never use liquid coolant on the band saw as coolant could damage the blade guide bearings or rubber tires on the pulleys. Blade wax may be used when cutting aluminum, brass, or thicker materials. Cast iron and steel should be cut dry.
10. Clamp all material firmly in a vise or by other clamping methods to prevent breaking blades or injury to the operator. Never attempt to saw material being held by someone.
11. Cut angle iron with legs down and rectangular material with widest side toward the blade.
12. To prevent pinching or twisting the saw blade or dropping the waste material on someone, support the waste material with a roller stand.
13. Mark the material to be cut and always saw on the waste side of the line.
14. Check the electrical cord to see that it is in good condition and use only electrically grounded outlets.
15. As the operator, assume a position of good balance and comfort for holding and controlling the saw.
16. Grip the saw by the front and rear handles making sure the trigger switch remains in the “off” position until ready to begin the saw cut.

17. Position the saw so the saw work stop is against the material being cut. If not, the saw will quickly pull itself back against the stop as it is started due to the direction of blade rotation. This quick movement could damage the blade or jerk the saw from the operator's hands.
18. Hold the saw so the blade is just above the material before depressing the trigger switch. Depress the switch and lower the blade gently onto the material.
19. Do not force the saw into the material. Hold the saw squarely and firmly against the material allowing the saw to do its job at its own pace with only the weight of the saw exerting pressure down on the material. Additional pressure will slow down the speed of the blade and reduce cutting efficiency. Do not rock the saw in an attempt to get it to saw faster as this will shorten blade life.
20. Do not attempt to break off the material before the cut is complete.
21. Hold firmly onto the saw as the cut is nearing completion to keep it from dropping or hitting the material as the cut is completed.
22. Release the trigger switch and allow the saw to stop before setting down on the bench or other safe resting area. Disconnect the saw from the energy supply.
23. Release the metal from the vise, clean the area, and remove all scrap from the floor.
24. Clean the saw and return it to its proper storage when the job is completed.

GENERAL SAFETY PRACTICES:

1. Wear approved eye protection, hearing protection, and proper clothing. Tie up loose hair and remove loose jewelry.
2. Do not operate the machine without the instructor's permission, or without instructor supervision.
3. The operator should always have both hands on the saw and make sure helpers are at a safe distance from the saw blade.
4. Keep the adjustable blade guide as close as possible to the material being cut.
5. Keep blade guards in place at all times.
6. Secure metal material to be cut in a vise or other clamping methods. Never attempt to saw material being held by a helper.
7. If a blade breaks, do not attempt to stop the saw. Release the trigger switch, and it will stop on its own.
8. The saw should be disconnected from the electrical power source before changing blades or making any adjustments.
9. Make sure the saw is properly grounded and that it is connected to a grounded electrical source having sufficient amperage for the saw.
10. Start the saw before the blade touches the material to be cut. Make sure the work stop is touching the material.

11. Do not force the speed of cutting by applying pressure to the saw other than the weight of the saw.
12. Support long material. Do not saw between two supports causing the material to bind or pinch the saw blade.
13. Remove burrs left by the saw on the material by filing or grinding.
14. Use a brush, not bare hands, to remove iron filings from the saw or work area.
15. When the job is completed, remove all scrap materials from the floor.

COMPLETION QUESTIONS:

1. When replacing the blade in the portable metal cutting band saw, the teeth should be pointing toward the ____________________________.
2. When cutting material that is 1/2" thick, a blade having ____________________________ teeth per inch should be selected.
3. The general rule for blade speed is the harder the material being cut, the ____________________ the blade speed, so when cutting carbon steel, the blade speed should be ____________________ to ____________________ SFPM.
4. A new blade could be damaged if allowed to enter a partially sawed material because the saw ________________________ is ________________________ than the new blade will cut.
5. If cutting high speed steel, select a blade speed of ____________________ SFPM.
6. Angle iron should be cut with the angle legs ____________________________.
7. The ____________________ ____________________ handle is used to release pressure on the blade to replace a worn blade.
8. Mark the material to be cut and always saw on the ______________________ side.
9. The ____________________ ____________________ should be held firmly against the material before depressing trigger switch to avoid the saw from being jerked from the operator's hands.
10. The saw should be ________________________ before the blade is allowed to touch the material.

PORTABLE POWER PLANER

PART IDENTIFICATION

Identify the numbered parts of the portable power planer illustrated below.

1. ____________________
2. ____________________
3. ____________________
4. ____________________
5. ____________________
6. ____________________
7. ____________________
8. ____________________

SAFE OPERATIONAL PROCEDURES

1. Study the operation, maintenance, and safety manual for the specific power planer to be operated.
2. Keep the knives sharp and properly adjusted.
3. When knives are replaced, they should be sharpened and replaced as a set. Both knives should weigh the same to maintain balance of the cutter head.
4. The depth of cut is adjusted by raising or lowering the front shoe of the planer. Start with a very thin cut, 1/16” or less.
5. The fence should be used when planing the edge of a piece of material. Set the fence perpendicular to the bottom of the plane to get a square edge.
6. The fence must be removed when planing flat surfaces.
7. When starting a cut, apply downward pressure on the front edge shoe and move forward slowly. Control the planer’s forward movement, keeping it slow and uniform.
8. To finish a cut, put more downward pressure on the rear shoe as the planer is moved forward over the material.
9. Plane with the grain, to avoid chipping.

GENERAL SAFETY PRACTICES

1. Wear approved eye protection, hearing protection, and proper clothing. Tie up loose hair and remove loose jewelry.
2. Do not operate the machine without the instructor's permission, or without instructor supervision.
3. Be sure the chip deflector is properly attached to the machine so the chips will fly away from the operator.
4. Work on clean material; nails and paint will damage the planer blades.
5. Be sure the work area is clean and free of scraps, sawdust, or chips.
6. Be sure the material is properly secured so it will not move while the material is being planed.
7. Disconnect the planer from the energy source when replacing or adjusting the blades or making any adjustments to the planer.
8. Do not make adjustments while the motor is running.
9. Never set the machine down until the motor has stopped.
10. Do not talk to anyone while using the planer.
11. Hold the machine with one hand on the handle and the other on the knob, if one is available. Be careful not to come in contact with the knives.
12. Keep the power cord away from the knives.
13. After use, disconnect the energy source, clean the planer, and return it to proper storage.

COMPLETION QUESTIONS

1. To make a uniform cut at the start of the material, apply the most pressure on the ____________________ ____________________ of the plane.
2. The ____________________ must be removed when planing flat surfaces.
3. The depth of cut is adjusted by raising or lowering the ____________________ ____________________ of the planer.
4. The ____________________ ____________________ prevents the chips from flying into the operator's face.
5. Before planing, always ____________________ the board.
6. The fence is used to help make a ____________________ cut on the edge of the board.
7. Plane ____________________ the grain when possible to avoid chipping the grain.
8. The depth of cut adjustment must be made while the motor is ____________________.
9. All knives should weigh the same to maintain ____________________ of the cutter head.
10. Start with a thin cut ____________________ or less.

PRESSURE WASHER

PART IDENTIFICATION

Identify the circled parts on the pressure washer illustrated below.

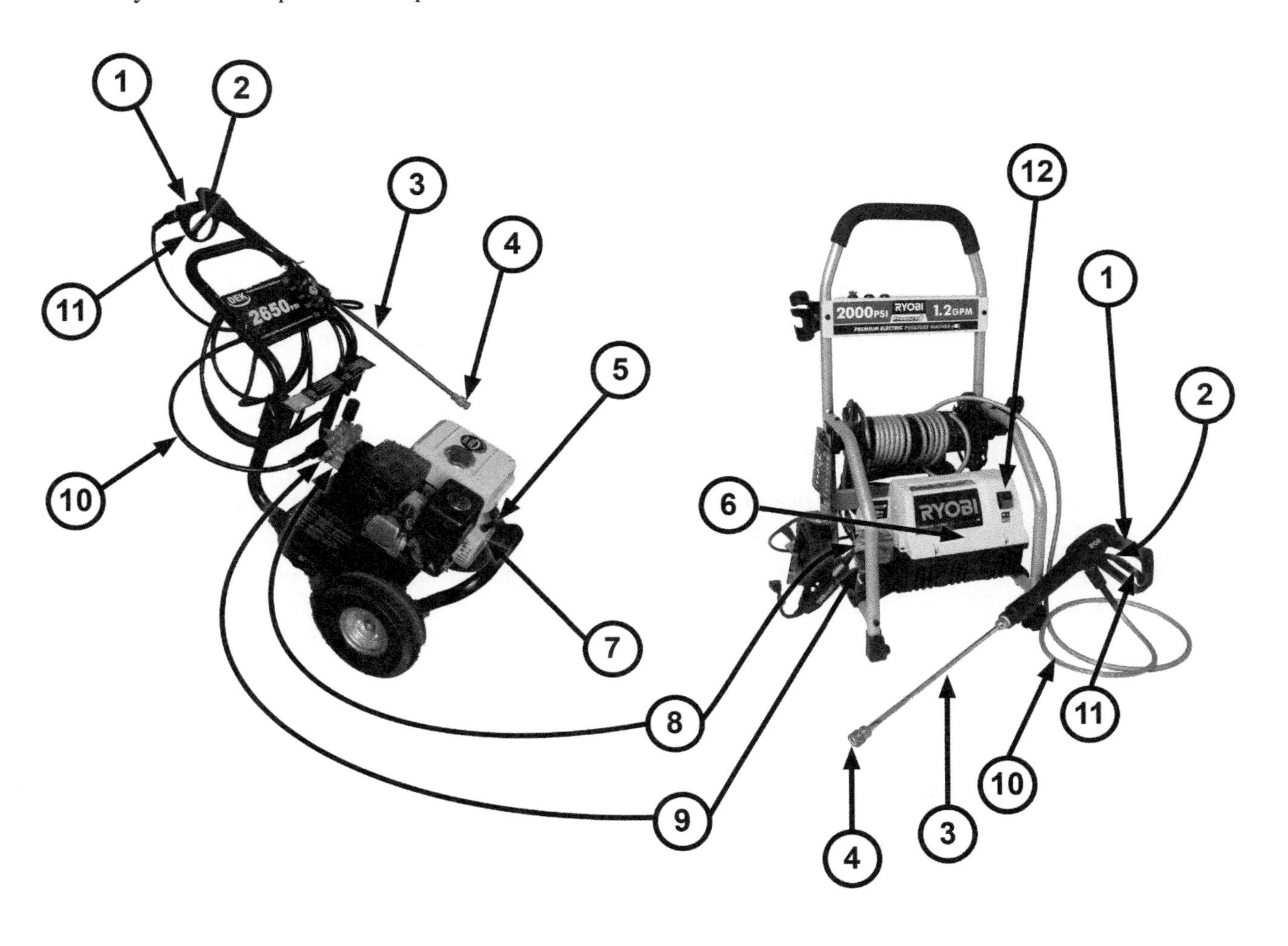

1. ______________________________
2. ______________________________
3. ______________________________
4. ______________________________
5. ______________________________
6. ______________________________
7. ______________________________
8. ______________________________
9. ______________________________
10. ______________________________
11. ______________________________
12. ______________________________

SAFE OPERATIONAL PROCEDURES

1. Start-Up
    a. Sweep or brush all loose dirt and debris from the surface you're cleaning.
    b. Connect a hose to the water inlet. Make sure your hose can supply the required GPM for the machine. Make sure both hoses are free of clogs and kinks.
    c. Fill the detergent reservoir (if using a cleaner) with a detergent solution.
    d. Set the spray wand to off or a low-pressure setting to prevent kickback when the washer is started. Washers with variable nozzles should be on a low-pressure, wide-angle setting. At this point, make sure there's no tip on a washer that uses interchangeable tips.
    e. Turn the faucet on fully.
    f. Squeeze wand trigger to release air pressure.
    g. Check hose connections for leaks.
    h. To start the engine or motor: For a gas power washer, adjust the choke and throttle if needed, brace your foot against a wheel to steady the machine, then pull the starter cord to start the engine. For electric, plug into a grounded, GFCI outlet and flip the on/off switch to "on".
2. Use
    a. Let water run through the machine for about a minute to prime the system. But never let it idle for longer than 3-5 minutes to prevent overheating the pump.
    b. With trigger off, attach a low-pressure nozzle, or keep adjustable nozzle on low. Always start with low pressure and work your way up.
    c. Hold the wand with both hands. Plant feet firmly on the ground.
    d. Start with low pressure, about 18 inches from cleaning surface. Move in a side-to side motion at a 45-degree angle to the surface.
    e. Keep moving. Don't stay in one spot or you risk damage to the surface. Move the spray in overlapping lines to avoid streaks.
    f. If you need more pressure, try moving the spray closer to the surface, but no more than 6 inches away. Or lock the trigger on the wand and attach a higher-pressure tip.
3. Shut-Down
    a. Release the trigger lock if used, and then wand trigger.
    b. Turn power off to machine.
    c. Turn off water supply to machine.
    d. Pull the trigger once more to release the water pressure before disassembling.

GENERAL SAFETY PRACTICES

1. Wear approved eye protection, hearing protection, and proper clothing. Tie up loose hair and remove loose jewelry.
2. Do not operate the machine without the instructor's permission, or without instructor supervision.

3. Be sure you always wear closed-toe shoes or boots when pressure washing. Wearing safety-toe boots is recommended, as high-pressure streams can potentially tear through rubber boots.
4. Gloves are one of the best forms of safety gear you can wear. They not only protect your hands but improve your grip, helping to avoid other unnecessary injuries.
5. Make sure your legs are protected from flying debris that could break skin. In addition, if you're cleaning sidewalks or pavement, you want to protect your legs from the downward jet of water that's near them.
6. Never aim the pressure washer at another person.
7. Use the safety latch when not spraying, to prevent unintentional engagement of the pressure washer.
8. Never use a gas pressure washer in enclosed areas.
9. Never attempt to push or move objects with spray from the washer.
10. Always test the ground fault circuit interrupter (circuit breaker or outlet) before using a pressure washer.
11. If an extension cord must be used, keep the pressure washer's power cord connection out of any standing water, and use a heavy-duty extension cord with components rated for use in wet locations.
12. Keep both the power cord and extension cord connections as far away as possible from the item being washed and away from any water runoff.

COMPLETION QUESTIONS

1. Check hose connections for ________________.
2. Never ________________ the pressure washer at another person.
3. Let water run through the machine for about a ________________ to prime the system
4. Never use a________________ pressure washer in enclosed areas.
5. Be sure you always wear ________________ shoes or boots when pressure washing.
6. Move the spray in ________________ lines to avoid streaks.
7. Sweep or brush all loose ________________ ________________ ________________ from the surface you're cleaning.
8. Use a ________________ extension cord with components rated for use in outdoor locations.
9. Use the________________ ________________ when not spraying, to prevent unintentional engagement of the pressure washer.
10. Hold the wand with ________________ ________________. Plant feet firmly on the ground.

RECIPROCATING SAW

PART IDENTIFICATION

Identify the numbered parts of the reciprocating saw illustrated below.

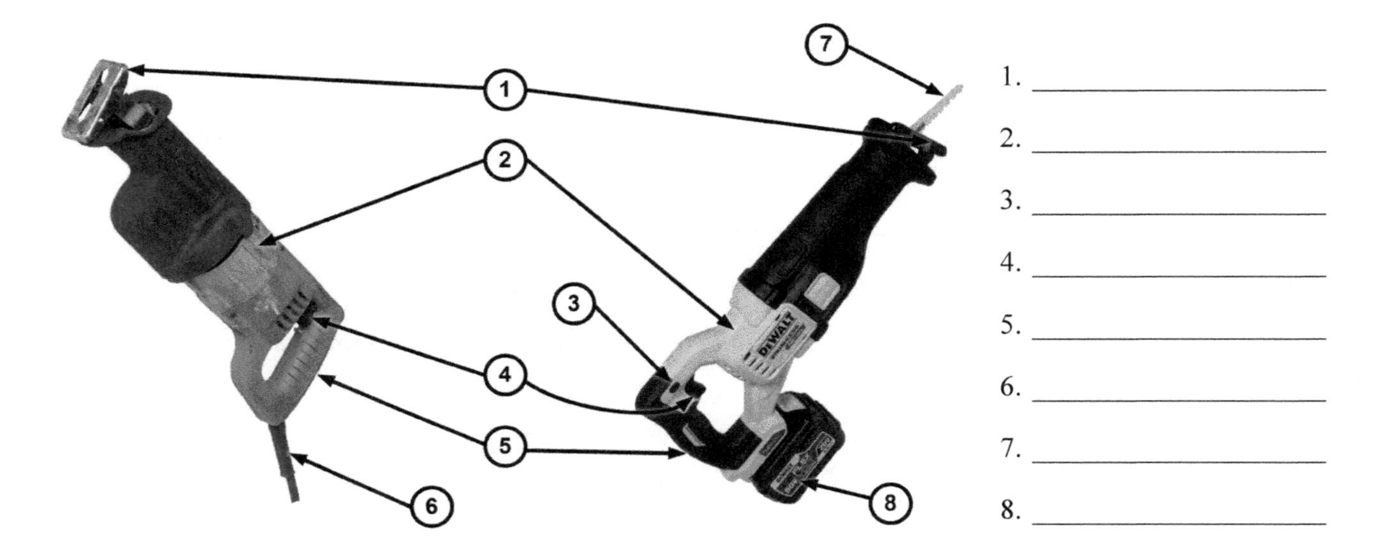

SAFE OPERATIONAL PROCEDURES

1. The reciprocating saw is a handheld power tool designed to make cuts in any direction thus making it more versatile than the jig saw. The blade moves in a straight line with the saw which makes for a simpler design and a better power factor than the right-angle movement of the blade in the jig saw.

2. Commonly used for demolition and remodeling, the reciprocating saw can be used by many and varied construction workers such as carpenters, electricians, and plumbers.

3. Most reciprocating saws have variable speed triggers/dials, which means you have the ability to control how fast the blade cuts. Reciprocating saws vary from 3,000-0 Strokes Per Minute (SPM). There is no universal speed setting which is designed for all materials and blade options. Generally, a high-speed blade is used for fast cutting or softer materials such as wood and plastics. A slow speed blade is generally used for more precise cutting, or for harder materials such as sheet metal, conduit, and pipe to prevent the blade from overheating.

4. Reciprocating saw blades may be installed with teeth facing up or down, depending on cutting requirements.

5. The length of stroke of the blade is usually ½" to 1¼". The longer stroke the saw makes, the faster it will cut. The shorter the length of stroke, the easier it will be to control.

6. Some reciprocating saws have a rocker type shoe which helps to roll the blade into starting cuts.

7. Blades for the reciprocating saw vary greatly in sizes, shapes, and types. Blades vary in length from 3”-12” and are made of high-speed steel and Bi-metal. High speed steel blades are used to cut hardwoods like oak, non-ferrous metal (metal that does not contain iron) like aluminum and copper, steel, reinforced plastic, and chipboard. A bi-metal blade consists of two different metals, high speed steel welded to a carbon steel backing. The high-speed steel provides a good cutting edge, while the carbon steel provides flexibility. Bi-metal blades are suitable for both metal and wood cutting applications. In general, woodworking blades are 6 to 12 teeth per inch. Blades with fewer teeth cut faster but leave a rough edge. Blades with more teeth cut slower but leave a smoother edge. When cutting metal, use a blade with at least 18 teeth per inch (TPI).

8. Blade length is determined by taking the width of material being cut and adding three inches. For example, if material being cut is 6” in width, the correct blade length would be 9”.

9. When you are using a reciprocating saw, you ideally should have three teeth in contact with the material at all times. When there are more than three teeth in the work, cutting performance declines, since waste can’t be expelled quickly enough. A blade with fewer teeth in the work is prone to damage because the working portion of the blade can become overheated.

10. Whatever the task to be done, the blade should be sharp, of proper type and length, correct teeth per inch (TPI), and correctly attached to the saw.

11. Making a plunge cut:
    a. Rest the saw on its shoe with the blade in line with the mark of the opening to be cut, yet not touching the material.
    b. Pull the trigger switch allowing the blade to enter the material. Slowly raise the angle of the saw until the blade enters the material.
    c. The saw may then be held at any angle that is convenient for the operator.
    d. When a corner is reached, back up the saw and make several cuts to the corner line.
    e. As soon as enough material has been removed, turn the saw and cut along the new line.
    f. A curved plunged cut is similar except the starting cut should be made to the inside of the outline so as not to make the curved edge oversize. Also, the saw must be held at a right angle to the material when cutting.

12. Making a straight or curved cut:
    a. In straight line cutting, set the shoe of the saw on the work.
    b. Start the motor and move the saw into the work but do not force the cutting action.
    c. Keep the shoe against the work at all times; the blade cuts from the bottom side to the top side, just opposite the cutting action of a handsaw.
    d. The good or finish side should be placed face down during the cutting operation.
    e. As with the curved plunge cut, when sawing curves, the saw must be held at a right angle to the work.

13. A flush cut (cut close up against a wall) can be made with most reciprocating saws by simply adjusting the blade carrier (horizontal) so it will clear the body of the saw. Follow similar procedures as in making opening cuts.

14. The reciprocating saw can be used for cutting metal by using the correct blade and following the same general procedures as for sawing wood. One important point is to be certain the metal is secured, either as constructed or held in a vise.
15. The saw should be kept clean, well lubricated, and periodically checked for loose parts and screws.
16. The blade should be removed when storing the saw.

GENERAL SAFETY PRACTICES

1. Wear approved eye protection, hearing protection, and proper clothing. Tie up loose hair and remove loose jewelry.
2. Do not operate the machine without the instructor's permission, or without instructor supervision.
3. If it is not double insulated, make sure the saw is properly grounded.
4. Disconnect reciprocating saw from energy supply when not in use, before servicing and cleaning, and when changing blades.
5. Make sure the shoe is always against the work piece.
6. Always use sharp, proper type, and size blade for the job being done.
7. Do not force the cut or attempt to turn too short a corner.
8. If using in an awkward position, such as above the head or on a ladder, make sure the operator and the material being cut are properly and securely supported.

COMPLETION QUESTIONS

1. The reciprocating saw is more versatile than the ____________________ saw because it cuts in a ____________________ ____________________ rather than at right angles to the saw base.
2. Most reciprocating saws have ________________________ __________________________ ____________________________ which means you have the ability to control how fast the blade cuts.
3. Reciprocating saw blades may be installed with teeth facing ________________________ or ____________________________.
4. The longer stroke the saw makes the ____________________ it will cut
5. Blades vary in length from ____________________ to ____________________ inches.
6. In general, woodworking blades are ____________________ to ____________________ teeth per inch
7. When cutting metal, use a blade with at least ____________________ teeth per inch.

8. Blade length is determined by adding ____________________ inches to width of material being cut.

9. You should have ____________________ teeth in contact with material at all times.

10. Whatever the task to be done, the blade should be ____________________, of proper type and ____________________, correct ____________________, and correctly attached to the saw.

ROTARY SAW/CUT OUT TOOL

PART IDENTIFICATION

Identify the circled parts on the rotary saw/cut out tool illustrated below.

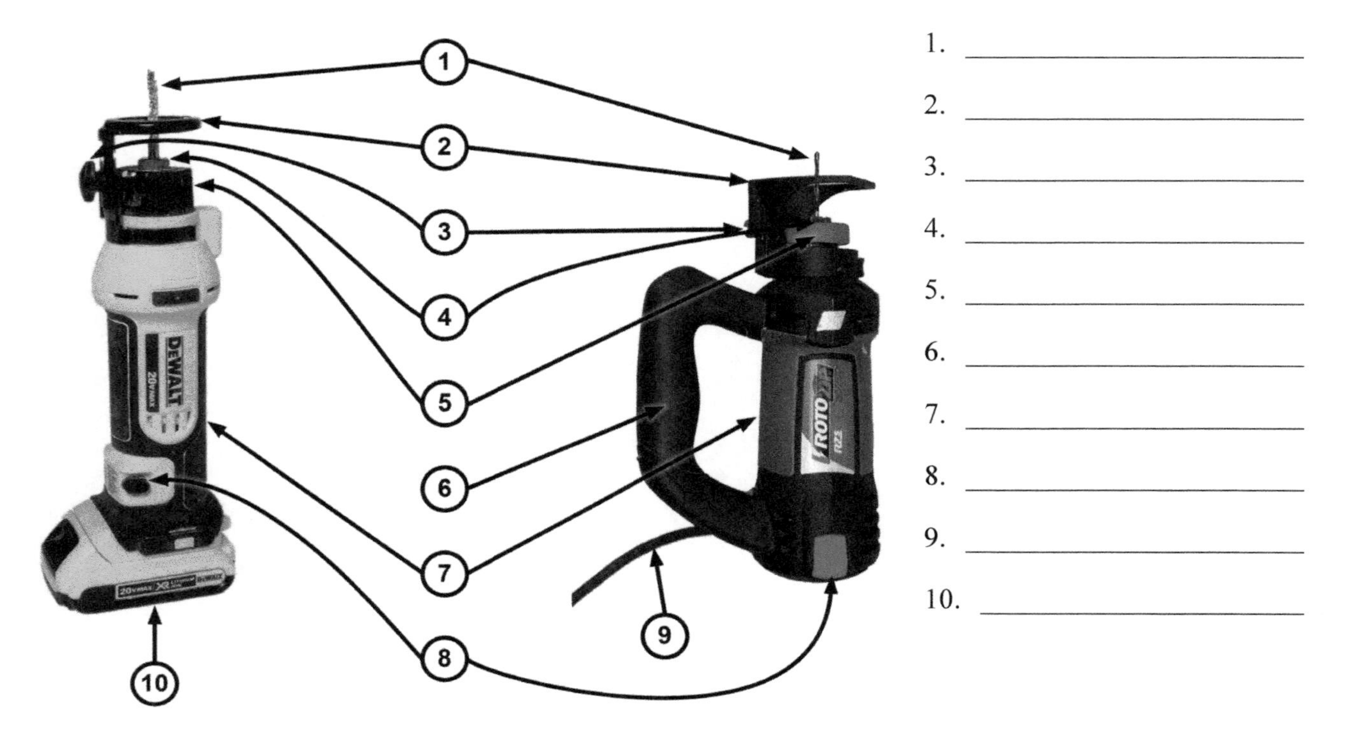

SAFE OPERATIONAL PROCEDURES

1. A rotary saw, or cut out tool is a type of mechanically powered saw used for making accurate cuts without the need for a pilot hole in drywall (gypsum board), plywood, or other thin, solid material.
2. Bit installation: bits are held in place by a keyless collet system or a bit retaining screw designed specifically for spiral saw bits with 1/8", 1/4", or 5/32" shanks.
3. For a keyless collet system follow these steps:
    a. Verify power source is disconnect and switch is in the off position.
    b. Depress and hold the shaft-lock and rotate the keyless chuck and shaft until the shaft-lock engages and holds the shaft.
    c. Rotate the keyless chuck (counter-clockwise), remove the old bit (if there is one) insert the new bit as far as possible, but not so far that the bit flutes are in contact with the jaws of the chuck. Leave approximately 1/8" of shank exposed.
    d. Re-engage the shaft-lock and securely tighten the keyless chuck (clockwise) by hand.
    e. When using the ¼" or 5/32" bits it may be necessary to use a wrench on the front of the keyless chuck to securely tighten the bit.

f. Verify that the depth guide bracket is securely attached to the collar by the lock lever.

g. Use the depth guide to adjust the depth of the cut by loosening the depth guide locking screw enough to enable the depth guide to slide up or down. The desired depth of cut is about 1/8” greater than the material thickness. Retighten the depth guide locking screw after depth guide is properly adjusted.

4. For a bit retaining screw system follow these steps:

   a. Verify power source is disconnect and switch is on the off position.

   b. Depress and hold the shaft-lock and rotate the shaft until the shaft-lock engages and holds the shaft.

   c. For 1/4” and larger bits. Use the hex wrench to loosen the bit retaining screw by turning counter-clockwise, remove the old bit (if there is one) insert the new bit at least 1/2” into the spindle. Leave approximately 1/8” of shank exposed.

   d. Several models require the use of a special chuck to properly secure a 1/8” shank bit. Refer to your owner’s manual for additional information.

   e. Re-engage the shaft-lock and securely tighten the bit retaining screw by turning clockwise.

   f. Verify that the depth guide is securely attached to the tool by the set screw.

   g. Use the depth guide to adjust the depth of the cut by loosening the depth guide set screw with a hex wrench enough to enable the depth guide to slide up or down. The desired depth of cut is about 1/8” greater than the material thickness. Retighten the depth guide set screw after depth guide is properly adjusted.

5. Making drywall cutouts around electrical boxes

   a. Verify that the box or fixture that requires cutout is firmly attached to framing and any electrical wires are pushed back and out of the way. The drywall cut is made on the outside edges of the box or fixture and is used as the guide. Do not attempt to make cut-outs around any fixture or electrical box which has live (energized) electrical wires.

   b. Mark the center location of the electrical box or fixture on drywall before installing.

   c. Make certain that the collet nut or bit retaining screw is securely tightened and depth guide is properly set before turning on the tool.

   d. Connect to the energy source, hold the tool firmly with both hands and turn the tool on.

   e. Plunge the bit through the mark you made, then guide the bit to the right until you feel and hear the bit touch the inside edge of the box.

   f. Pull the bit out far enough to slip it over the edge of the box so it is now against the outside of the box.

   g. While keeping the bit in contact with the outside of the box, move the tool counter-clockwise while applying slight inward (towards center of box) and upward pressure. Continue moving the tool counter-clockwise against the outer edge of the box until you have completed the cut.

   h. After completing your cut, turn off the tool and carefully remove it from the material.

6. Making cut-out in material other than drywall.

a. Make certain that the collet nut or bit retaining screw is securely tightened and depth guide is properly set before turning on the tool.
b. Connect to the energy source, hold the tool firmly with both hands and turn the tool on.
c. While holding the tool firmly, insert the bit into the material at a 45° angle.
d. Slowly bring it to a 90° angle to begin the cut.
e. Because of the rotating cutting action, there will be a slight pull when cutting. The slower you cut, the more control you have.
f. Excessive pressure or fast cutting will cause excessive heat and may shorten the life or break the cutting bit.
g. The base guide should be flush to the material surface. For all materials (except cutting around electrical boxes in drywall), steer the tool in a clockwise direction with slow, steady pressure to make the cut following your mark.
h. When cutting on a vertical surface, avoid ending your cut at the bottom of your opening. If possible, start and end at the top so the scrap part will not drop onto the rotating bit.
i. After completing your cut, turn off the tool and carefully remove it from the material.

GENERAL SAFETY PRACTICES

1. Wear approved eye protection, hearing protection, and proper clothing. Tie up loose hair and remove loose jewelry.
2. Do not operate the tool without the instructor's permission, or without instructor supervision.
3. Dust mask must be used for appropriate conditions.
4. If using a corded tool, inspect that the prongs are secure and the cord has no cuts or exposed wires before use.
5. Hold tool by insulated gripping surfaces when operating the tool and always with two hands during start-up.
6. Avoid accidental starting. Be sure switch is "off" before connecting to power source.
7. If applicable, remove adjusting keys or wrenches before turning on the tool.
8. Always make sure the work surface is free from nails and other foreign objects.
9. Never hold the workpiece in one hand and the tool in the other hand when in use. Clamp the material and guide the tool with both hands.
10. Never lay workpieces on top of hard surfaces like concrete, stone, etc. Protruding cutting bit may cause tool to jump.
11. After changing the bits, or making any adjustments, make sure the collet nut or bit retaining screw is securely tightened.
12. Never start the tool when the bit is engaged in the workpiece.
13. Never use dull or damaged bits, they can snap during use. Sharp bits must be handled with care.

14. Never touch the bit during or immediately after use. After use the bit is too hot to be touched by bare hands.
15. Never lay the tool down until the motor has come to a complete standstill.
16. Do not use the tool for drilling purposes.
17. Always use the tool with the depth guide securely attached and positioned flat against material being cut.
18. Do not use accessories that are not designed and recommended by the manufacturer.

COMPLETION QUESTIONS

1. Bits are held in place by a keyless collet system or a bit _______________ screw.
2. The desired depth of cut is about _______________ greater than the material thickness.
3. The drywall cut is made on the _______________ edges of the box or fixture.
4. While holding the tool firmly, insert the bit into the material at a _______________ angle.
5. When cutting on a vertical surface, avoid ending your cut at the _______________ of your opening.
6. Tie up loose hair and remove _______________ jewelry.
7. Always make sure the work surface is _______________ from nails and other foreign objects.
8. Never start the tool when the _______________ is engaged in the workpiece.
9. Never lay the tool down until the motor has come to a _______________ standstill.
10. Do not use accessories that are not designed and _______________ by the manufacturer.

ROTARY TOOL

PART IDENTIFICATION

Identify the circled parts on the rotary tool illustrated below.

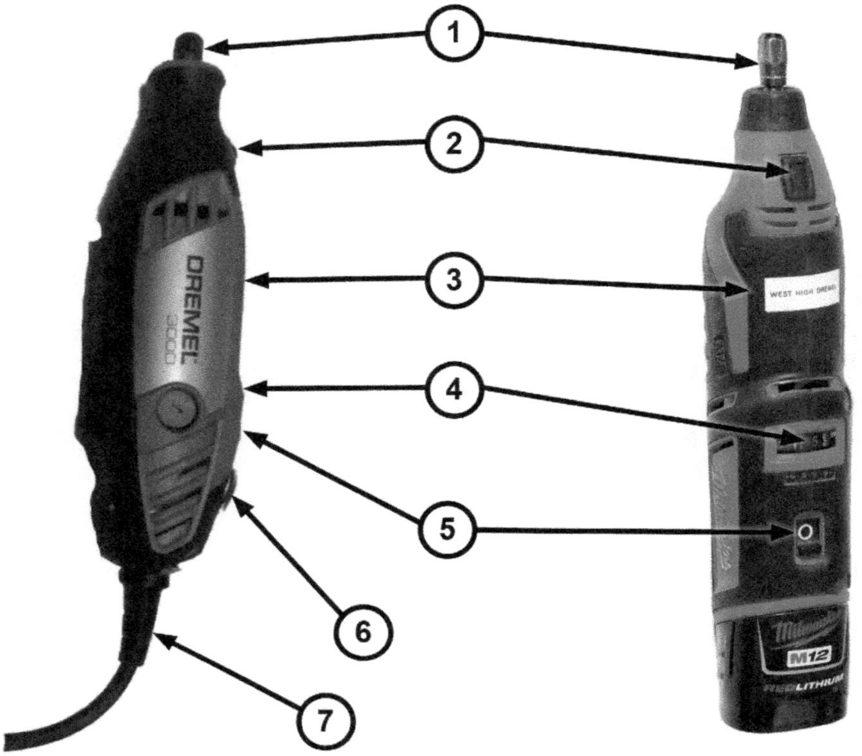

1. ____________________
2. ____________________
3. ____________________
4. ____________________
5. ____________________
6. ____________________
7. ____________________

SAFE OPERATIONAL PROCEDURES

1. Secure the material on a table or stand before starting to operate the tool.
2. Fitting the bit/accessory.
   a. Before fitting the bit/accessory, ensure that the tool is disconnected from the power source.
   b. Turn the collet nut slowly to loosen the collet, whilst pressing the spindle lock/button.
   c. The spindle lock button will engage, and prevent the collet from rotating.
   d. Keeping the lock down button depressed, turn the collet nut to loosen the collet.
   e. Insert the required bit/accessory into the collet.
   f. Clamp all attachments as short as possible. Shafts that stick out too far may bend or cause the tool to vibrate.

g. Tighten the collet nut again depressing the lock down button. Ensure that the bit/accessory is firmly secured in the collet and the collet nut is tightened firmly.

3. Bit/accessories requiring extra setup

   a. Sanding Disc Arbor

      (1) To fit a disc to the Sanding Disc Arbor, unscrew the fitting screw.

      (2) Remove the screw and washer.

      (3) Place the required disc on the end of the arbor.

      (4) Replace the washer and screw.

      (5) Tighten the screw until the disc is firmly secured.

      (6) Fit the arbor to the tool as described in “Fitting the bit/accessory”.

   b. Sanding Drum Arbor

      (1) To fit a disc to the Sanding Drum Arbor, unscrew the fitting screw.

      (2) Remove the screw, washer and rubber drum holder.

      (3) Push the required Sanding Drum on to the rubber drum holder.

      (4) Ensure that the drum is completely pushed on to the holder so that edges of the drum and holder are aligned.

      (5) Replace the drum holder on the arbor.

      (6) Replace the washer and screw.

      (7) Tighten the screw until the disc is firmly secured.

      (8) Fit the arbor to the tool as described in “Fitting the bit/accessory”.

   c. Polishing Wheel Arbor

      (1) Push the threaded point of the Polishing Wheel Arbor into the required polishing wheel.

      (2) Twist the arbor counter-clockwise to secure the polishing wheel firmly on to the arbor.

      (3) Fit the arbor to the tool as described in “Fitting the bit/accessory”.

      (4) Occasionally check the wheel during use to ensure it is still secure.

4. Always ensure that the switch is in the off position before connecting the tool to the power source. Always disconnect the tool from the power source when not in use.

5. Press the switch into the on position.

6. Run the tool without a load to check the vibration level before using tool.

7. Adjusting the Speed

   a. The tool has a range of speed settings.

   b. As a guide, use low speed for larger attachments and higher speed for small attachments.

   c. Do not force the tool. Best performance is achieved by an even speed, not high pressure.

   d. High Speed Uses:

      (1) Higher speeds are better for carving, cutting, routing shaping. Hardwoods, metals and glass require high speed operation and drilling should also be done at high speeds.

(2) Do not use grinding wheels over 1" in diameter. The tool's high speed can cause larger wheels to fly apart and could cause injury.

e. Low Speed uses:

(1) Slow speeds usually are best for polishing operations employing the felt polishing accessories. They may also be for working on delicate projects.

8. Periodically stop the tool and check for wear on the bit/accessory.
9. Press the switch back into the off position.

GENERAL SAFETY PRACTICES

1. Wear approved eye protection, hearing protection, and proper clothing. Tie up loose hair and remove loose jewelry.
2. Do not operate the machine without the instructor's permission, or without instructor supervision.
3. Be sure no one is in the work area.
4. Use caution when replacing bit/accessory, they can be very hot. Use pliers or leather gloves to remove recently used bits.
5. Operate the tool to control dust and fumes, and have proper ventilation.
6. To clean the rotary tool when the job is complete.

COMPLETION QUESTIONS

1. _______________ _______________ should be used to protect eyes from flying fragments.
2. Remove any loose _______________ when you are operating the rotary tool.
3. Before you plug in the rotary tool, check the _____________ to be sure it is in the off position.
4. Do not force the tool, use _______________ speed for best performance, not heavy pressure.
5. The _______________ _______________ _______________ is used to hold the collet in place when changing bits and accessories.
6. Use caution when changing the _____________ or accessories; they will be hot after operation.
7. Do not operate the machine without the instructor's permission, or without instructor _______________.
8. Tie back your _______________ to prevent it from getting caught in the spinning tool.
9. Do not use grinding wheels over _______________ in diameter. The tool's high speed can cause larger wheels to fly apart and could cause injury.
10. Clamp all attachments as _______________ as possible.

ROUTER

PART IDENTIFICATION

Identify the numbered parts of the router illustrated below.

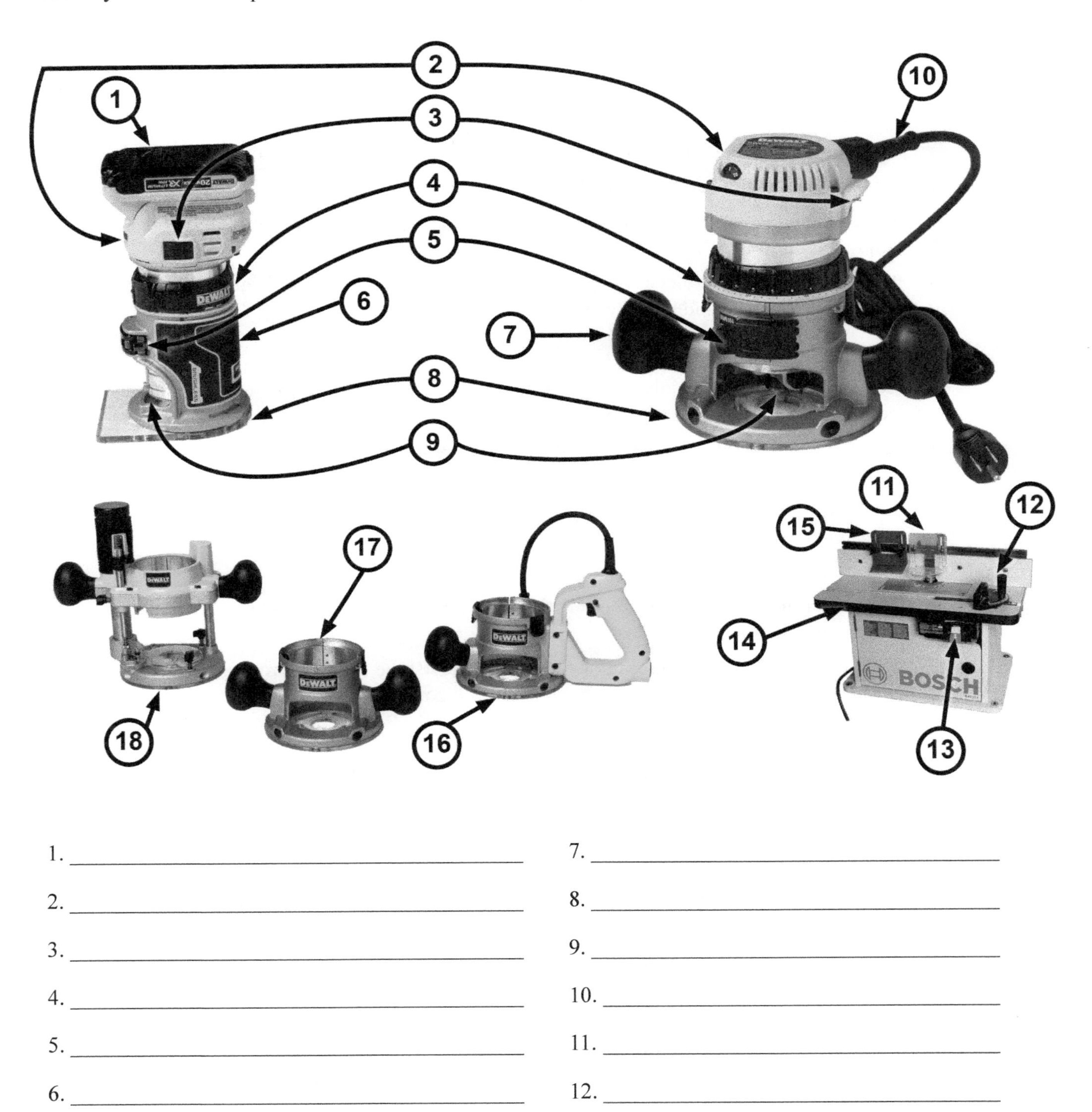

1. ______________________________

2. ______________________________

3. ______________________________

4. ______________________________

5. ______________________________

6. ______________________________

7. ______________________________

8. ______________________________

9. ______________________________

10. ______________________________

11. ______________________________

12. ______________________________

13. ______________________________ 16. ______________________________

14. ______________________________ 17. ______________________________

15. ______________________________ 18. ______________________________

SAFE OPERATIONAL PROCEDURES

1. Changing the router bits or cutters:
   a. Disconnect router from energy source.
   b. Select proper bit for job to be completed.
   c. Have proper router collet tools available for loosening and tightening the chuck.
   d. Loosen the locking handle and remove the router from the base.
   e. Check manufacturer's procedure for changing bits, in particular, for the method of holding motor or collet to properly tighten bit.
   f. Insert router bit into collet as deep as it can go and back it out 1/16". Tighten securely. Turn the router by hand to make sure the bit can turn freely.
2. Cutting a rabbet, dado, or molding:
   a. Select proper bit for the job. It must be sharp; look for chipped areas along the cutting edge of the bit. Replace router bits when they become chipped or damaged.
   b. Clamp work securely and make all adjustments before connecting the router to the energy source and starting the router.
   c. Lock the cutter in the router and adjust the base to the desired height using depth adjustment gauge. Lock the depth adjustment.
   d. If cutting a groove, a guide is required.
   e. Place the router base on the material with the bit clear of the material before turning on the power.
   f. Hold router firmly when turning on the motor to resist the starting torque. A well-balanced stance is important to help maintain full control.
   g. Make a practice cut on a piece of scrap material.
   h. Hold the router with both hands, using the handles provided.
   i. When making a cut along a straight edge, against the rotation, cut end grain first, then edge grain to avoid chipping off material at the ends.
   j. Use steady, slow, even feed. Don't overload the motor.
   k. Cut only clean material. Keep in mind: cutting plywood dulls the bit because of the glue in the plywood.
   l. When the cut is completed, move the router bit away from the material; turn off the power. Do not lift the router from the material until the motor and bit have come to a complete stop.
   m. Between cuts lay the router on its side on a table in a position where it will not roll or rest on the bit.

n. Disconnect energy source, then remove the bit from the router after the job is completed and return it to its proper storage place.

3. Operating a router table:
   a. Select proper bit for the job. It must be sharp; look for chipped areas along the cutting edge of the bit and replace router bits when they become chipped or damaged.
   b. Lock the bit in the router and adjust the router to the desired height using depth adjustment gauge. Lock the depth adjustment.
   c. If cutting a groove, a fence is required.
   d. Place material on the router table with the bit clear of the material before turning on the power.
   e. Make a practice cut on a piece of scrap material.
   f. Hold the material with both hands, keeping fingers/hands 4" away from cutting bit.
   g. When cutting start with the end grain first, then edge grain to avoid chipping off material at the ends.
   h. Use steady, slow, even feed. Don't overload the motor.
   i. Cut only clean material. Keep in mind: Cutting plywood dulls the bit because of the glue in the plywood.
   j. When the cut is completed, move the material away from the bit; turn off the power.

GENERAL SAFETY PRACTICES:

1. Wear approved eye protection, hearing protection, and proper clothing. Tie up loose hair and remove loose jewelry.
2. Do not operate the machine without the instructor's permission, or without instructor supervision.
3. Use only sharp cutters for the job to be done. Look for chipped areas along the cutting edge of the bit. Replace router bits when they become chipped or damaged.
4. Double check all adjustments to be certain they are tight.
5. Be sure the router is disconnected from the energy source when changing bits.
6. Never start the router when the router bit is in contact with the material.
7. Do not talk to anyone while operating the router.
8. Check to see if the switch is off before connecting to an energy source
9. Work area must be free of debris and obstructions. Routers create a lot of sawdust quickly, so pause and clean up throughout the process.

COMPLETION QUESTIONS:

1. The router is a dangerous power tool because the ________________________ is not guarded.

2. The router bit is held in a ________________________ style chuck.

3. The variety of router cuts is obtained from the great assortment of ______________________ available.

4. The bit should be inserted into the shank all the way and back it out ________________________ inch.

5. Keep fingers/hands ________________________ away from the moving bit.

6. The cutter will be dulled when cutting ________________________ because of the glue.

7. A router should be laid on its ________________________ between cuts.

8. A ________________________ or chipped bit should not be used.

9. Be sure the switch is in the ________________________ position before inserting the plug into an outlet.

10. When operating the router, the direction of travel should be ___________________________ the rotation of the bit.

SHIELDED METAL ARC WELDER (STICK)

PART IDENTIFICATION

Identify the numbered parts of the shielded metal arc welder (stick) illustrated below.

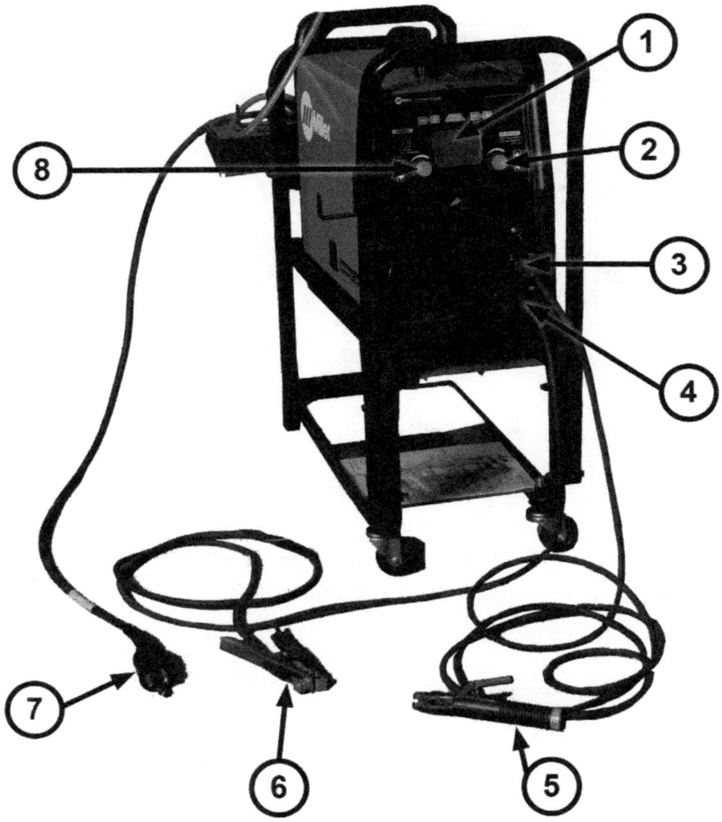

1. ____________________
2. ____________________
3. ____________________
4. ____________________
5. ____________________
6. ____________________
7. ____________________
8. ____________________

SAFE OPERATIONAL PROCEDURES

1. Electrode selection:
    a. The size of electrode to use is determined by the thickness of the metal. The most common sizes used are 1/8" and 5/32".
    b. The type of electrode is determined by the material to be welded, welding position, and the type of welder used.

Metal to be Welded	A.W.S. Number	Welder
Rusty, dirty steel	E6011	AC or DC-RP
Clean, mild steel	E6013	AC or DC-SP
Clean, mild steel	E6024	AC or DC-RP
Higher carbon steel	E7018	DC-RP or AC

c. American Welding Society (A.W.S.) numbers are given to electrodes so that electrodes made by different manufacturers can be compared on the same basis.

2. Adjusting the welder and striking the arc:

    a. Select the proper electrode considering the thickness and type of metal to be welded and the type of welder.

    b. Set the amperage by using the amperage chart given to you by the manufacturer of the machine you are using. Never change this while under load.

    c. Inspect electrode holders for defective jaws and poor insulation and keep all cable connections tight.

    d. Keep welding screens in place and turn fumes removal (ventilation) system on.

    e. Clear all combustible materials from the welding area before welding.

    f. Position a piece of metal on the welding table and check to make sure the welding machine is properly grounded. The welding equipment should be installed according to provision of the National Electric Code and the manufacturer's recommendations.

    g. Select the proper arc welding helmet and clothing to wear.

    h. Give the word "welding" or "cover" to all people standing nearby when ready to strike an arc, and protect others with the welding screen whenever possible.

    i. With everything in position, strike the arc as a match would be struck. Two other methods of starting the arc are by gently tapping the electrode to the metal or hovering slightly over it until the arc starts up.

    j. Raise the electrode about 1/2" from the metal as the arc is established; then slowly lower the electrode to about 1/8" or where a normal arc gap is maintained and a molten puddle is formed.

    k. If the electrode sticks, first attempt to twist it free from the metal. If this fails, release the holder from the electrode, turn off the switch, and use pliers to remove the electrode.

    l. Never turn off the switch while current is flowing through the electrode holder. This will arc the switch.

3. Welding a bead:

    a. Follow the steps in striking an arc.

    b. When the arc is established, tilt the electrode 15° to 20° in the direction of travel (from left to right for right-handed welders).

    c. Move the electrode ahead at a smooth and uniform rate.

d. Keep the electrode at the forward edge of the crater and feed it into the work as it melts off to maintain the proper arc length.
e. The width of the bead should be 1-1/2 times the width of the electrode.
f. When complete, chip the slag with a chipping hammer and always use a pliers when handling hot metal

GENERAL SAFETY PRACTICES

1. Wear approved eye protection, hearing protection, and proper clothing. Tie up loose hair and remove loose jewelry.
2. Do not operate the machine without the instructor's permission, or without instructor supervision.
3. Always wear a welding helmet with at least a number 10 lens when operating a shielded metal arc welder (stick). Never weld with a cracked shield or helmet and always avoid looking at the arc with the unprotected eye.
4. Wear long sleeves and gauntlet style (cover up to forearm) gloves. Wool or cotton clothing is best. DO NOT wear clothing made of synthetic fibers when welding. Some synthetic fibers are highly flammable. Avoid cuffs on trousers, pockets, and perforated or low-top shoes. High top leather shoes are best. A welding apron or leathers, and leggings could also be used.
5. Guard against the use of damp or wet clothing when welding. The use of such clothing increases the possibility of electrical shock.
6. Do not weld while wearing dirty or oil-soaked clothing.
7. Avoid exposure of any skin surface when operating a shielded metal arc welder (stick).
8. Wear eye protection at all times when operating a shielded metal arc welder (stick). Your welding helmet is not impact-resistant like proper eye protection is.
9. Perform work standing on concrete or fireproof surfaces and make sure the surface is not wet or damp.
10. Avoid welding directly on concrete floors. Residual moisture in the concrete may be turned to steam, resulting in the concrete exploding.
11. Do not weld in areas that store compressed gas cylinders.
12. Never weld on a container until it has been made safe.
13. Do not lay the electrode holder on a grounded table or surface. Place it on an insulated hanger. An electrode holder should never touch a compressed gas cylinder.
14. Place electrode stubs in a suitable container. Do not allow stubs to get on the floor in the welding area.
15. When making repairs or adjustments on the welder, always make sure the welder has been turned off and unplugged.

16. Clean the floor and table of scraps before starting to weld; keep welding tools in their proper location and keep your area free from water.
17. Make sure the work area is free of flammable, volatile, or explosive materials. (Ex. propane, gasoline, grease, and coal dust)
18. Do not carry matches, butane or propane lighters or other flammables in your pockets while welding.
19. Do not weld without proper ventilation.
20. The fumes given off by the following metals are toxic and may cause sickness or death: lead, cadmium, chromium, steel, manganese, brass, bronze, beryllium, zinc, or galvanized steel. DO NOT weld on these types of metal without obtaining permission from your instructor first. Welding may need to be done in an outdoor location.
21. Check welding cables for damaged insulation before welding. Do not use cables with frayed, cracked or bare spots in the insulation. keep them out of the way of sparks, hot metal, open flames, sharp edges, oil and grease while welding and make sure to wrap up or hang up all cables when done welding to prevent them from being damaged.
22. Never weld with the cables coiled over the shoulders.
23. Use caution in releasing the ground clamp after welding; it may be hot due to poor contact with the table.
24. Do not use water to extinguish fires near the welder. Use the dry chemical fire extinguisher or the welding blanket if you are extinguishing a fire on an individual.
25. Do not use a welder that does not have its case properly grounded.
26. Do not leave hot metal where others might come in contact with it.
27. Do not adjust the welder while it is in operation.
28. Treat all cuts or burns promptly. Report accidents to the instructor immediately.

COMPLETION QUESTIONS

1. A.W.S. stands for ____________________ ____________________ ____________________.
2. The proper arc length when welding is about ____________________ inch(es).
3. The arc is struck like striking a ____________________.
4. The angle the electrode is tilted in the direction of travel is ____________________° to ____________________°.
5. Two common sizes of electrodes used are ____________________ and ____________________.
6. The width of the bead should be ____________________ times the diameter of the electrode.

7. ______________________________ ______________________________ determines the size of electrode to use.

8. If the electrode sticks to the work, first attempt to ______________________________ it free.

9. Never ____________________ ____________________ the switch while current is flowing through the electrode holder.

10. While welding a bead, keep the electrode at the ______________________________ edge of the crater.

SOLDERING GUN

PART IDENTIFICATION

Identify the circled parts on the soldering gun illustrated below.

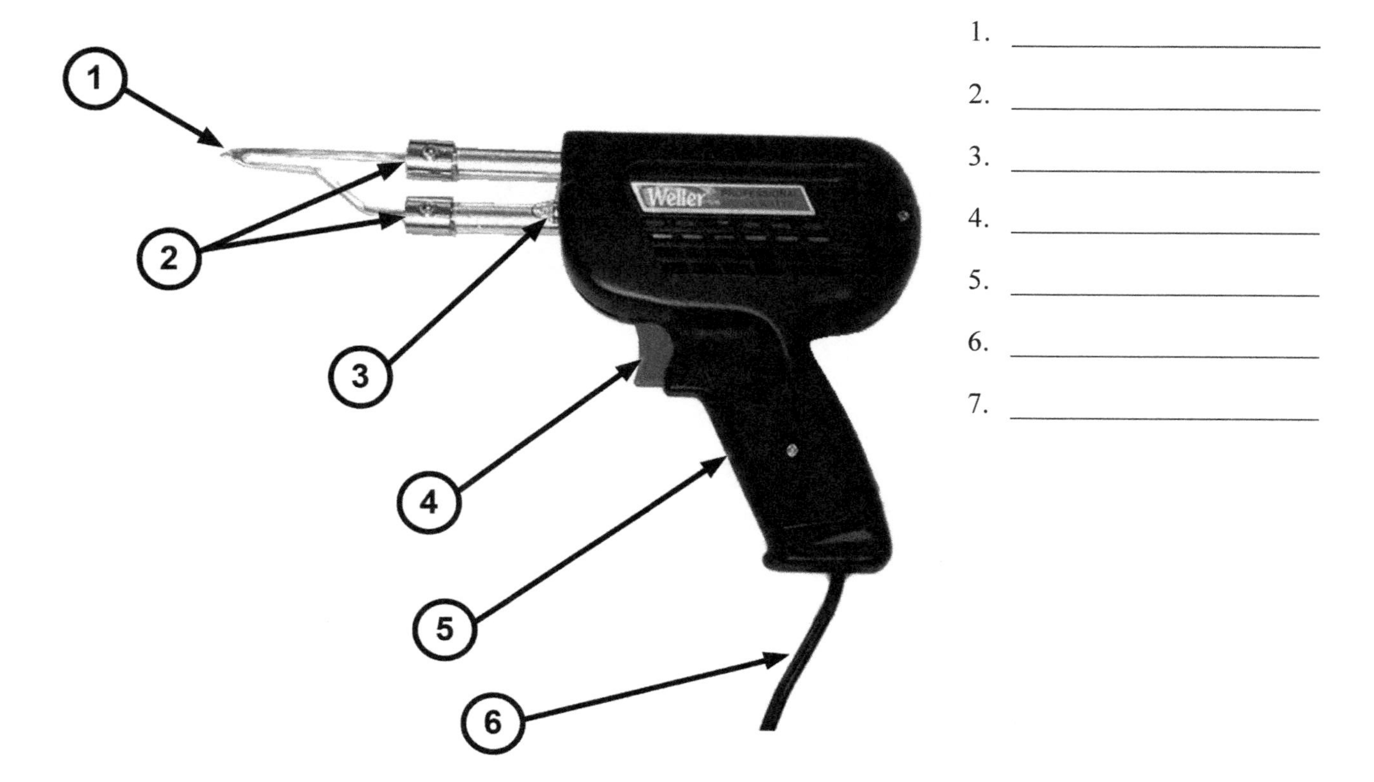

1. ____________________
2. ____________________
3. ____________________
4. ____________________
5. ____________________
6. ____________________
7. ____________________

SAFE OPERATIONAL PROCEDURES

1. Inspect the soldering gun, look over the cord for damage, ensure the tip is sharp and secure. Clean the cool tip with steel wool or fine emery cloth.
2. Prepare your soldering area. Remove all flammable materials and have a stand ready to rest the hot soldering gun on. Have a wet sponge to clean the hot soldering gun tip.
3. Prepare the materials you are soldering by determining how to hold the heated parts (clamps or pliers) and applying flux if needed. Prepare a straight length of solder off the spool.
4. Ensure all energy sources are turned off for items you are soldering.
5. When ready to solder, pull the trigger to heat the tip.
6. "Tin" the gun by applying a small amount of solder to the hot tip. Continue to tin the tip often to allow easy transfer of material.
7. Wipe excess solder off of the tip using a wet sponge or a special tip cleaner made from brass wire.

8. To solder, heat the connection first (not the solder) for a few seconds.
9. Leave the tip of the gun on the connection and add solder to the connection. The connection should get hot enough to melt the solder without touching it to the tip.
10. Only heat the connection enough to get the solder to flow. Do not overheat.
11. Allow the solder to flow through and around the connection before removing the iron.
12. Place the soldering gun in the stand to cool and allow the connection to cool completely before use.

GENERAL SAFETY PRACTICES

1. Wear approved eye protection, hearing protection, and proper clothing including long pants and sleeves to prevent burns. Tie up loose hair and remove ALL jewelry.
2. Do not operate the equipment without the instructor's permission, or without instructor supervision.
3. Read and understand the SDS for all materials before beginning work.
4. Only solder in areas with proper ventilation, benchtop ventilation may be necessary. 60/40 solder contains lead and the smoke from soldering should not be inhaled as it is toxic, gently blow the smoke away from your face. Lead-free solder is available.
5. Only hold the soldering gun by the handle, all other parts will be extremely hot—approximately 750° F.
6. Hold wires or other components to be heated with insulated tweezers or clamps.
7. Always place the soldering gun back in the stand, never pass it off to another person.
8. Do not blow on the solder to cool it down, it is molten metal and could spray, instead wait a few seconds before soldering other components.
9. Always complete work over a table/bench so molten solder cannot drip onto you.
10. Keep cleaning sponge wet during use.
11. Never secure the trigger down. Soldering guns are not rated for continuous use.
12. Always wash hands with soap and water after soldering.
13. Keep the soldering station free of electrical cables to prevent damage from the heated tip.
14. Know where the fire extinguishers are and how to operate them in case of emergency.
15. Seek medical attention if burned.
16. If using lead solder, sponges and other contaminated waste must be disposed of as hazardous waste.

COMPLETION QUESTIONS

1. By ________________ the soldering gun tip often it will allow the solder to transfer from the soldering gun to the material easily.
2. To see if the soldering gun is hot test the tinning process by touching ________________ to the gun tip.
3. Only solder in areas with proper ________________.
4. Hold wires or other components to be heated with ________________ tweezers or clamps.
5. Do not connect an ________________ ________________ to materials being soldered.
6. Wipe excess solder off of the tip using a wet ________________ or a brass wire tip cleaner.
7. 60/40 solder contains ________________ and the smoke from soldering should not be inhaled.
8. Allow the solder to flow ________________ and ________________ the connection before removing the gun.
9. Always complete work over a ________________ so molten solder cannot drip onto you.
10. Always ________________ your hands after soldering.

SOLDERING IRON (PEN) AND SOLDERING STATION

PART IDENTIFICATION

Identify the circled parts on the soldering iron (pen) and soldering station illustrated below.

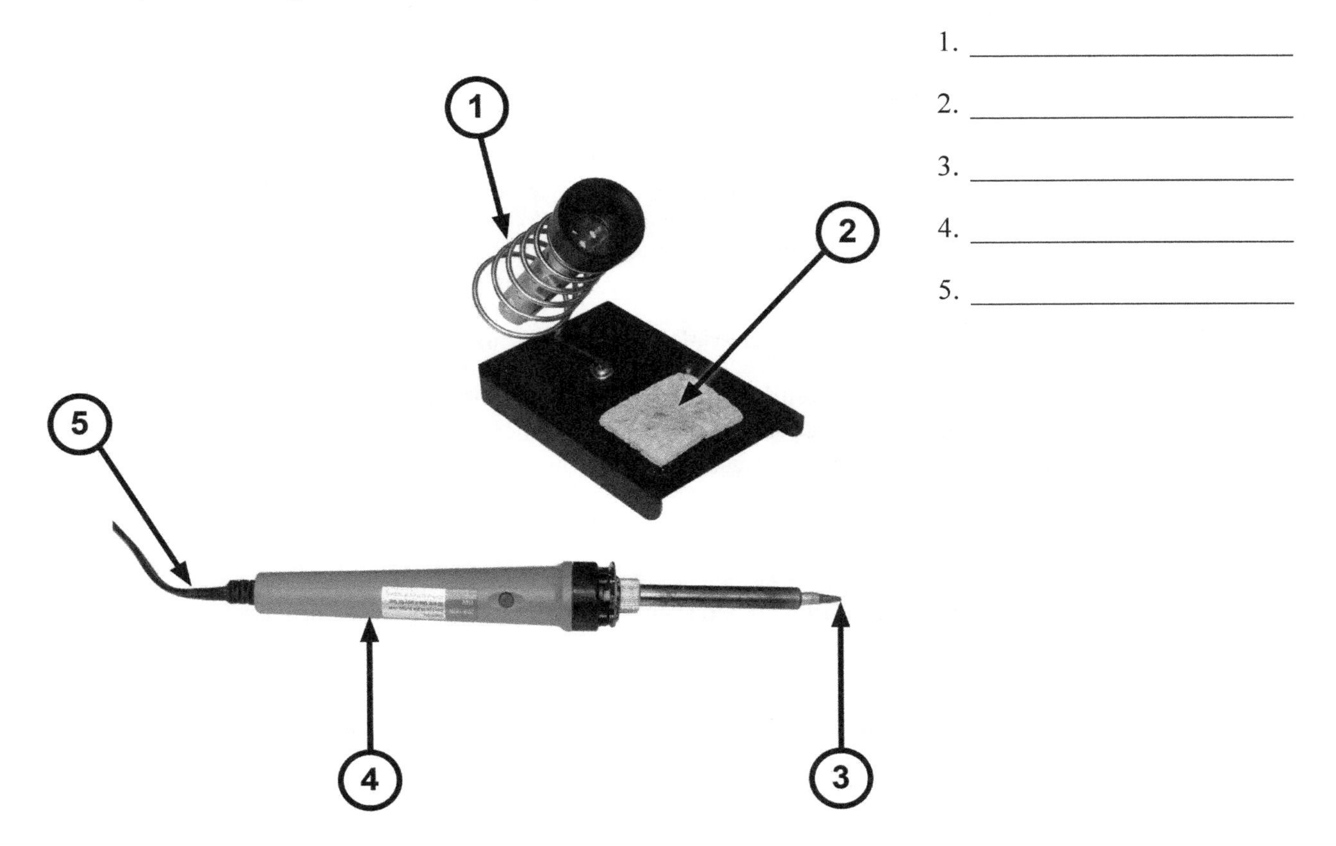

1. ____________________
2. ____________________
3. ____________________
4. ____________________
5. ____________________

SAFE OPERATIONAL PROCEDURES

1. Inspect the soldering iron, look over the cord for damage, ensure the tip is sharp and secure.
   a. After turning on (plugging in) the soldering iron, allow a few minutes for the iron to heat up.
   b. When the iron is hot enough the solder will melt on the tip, tin the tip before soldering.
   c. Wipe excess solder off of the tip using a wet sponge or a special tip cleaner made from brass wire.
   d. To solder, heat the connection first (not the solder) for a few seconds.
   e. Leave the iron in place and add solder to the connection.
   f. Allow the solder to flow through and around the connection before removing the iron.
   g. Don't apply too much solder to the connection.
   h. Do not overheat the connection or components, if necessary, move on to another component and return to the current component once it has had a chance to cool down.

2. Tin the soldering iron tip often to allow the solder to transfer from the soldering iron to the material easily.

3. Use a solder sucker or solder wick/ribbon to remove excess solder, or bridged connections which could create shorts amongst components.

4. Do not connect an energy source to materials being soldered, or materials recently soldered, all connections and components should be cool.

GENERAL SAFETY PRACTICES

1. Wear approved eye protection, hearing protection, and proper clothing including long pants and sleeves to prevent burns. Tie up loose hair and remove ALL jewelry.
2. Do not operate the equipment without the instructor's permission, or without instructor supervision.
3. Read and understand the SDS for all materials before beginning work.
4. Only Solder in areas with proper ventilation, benchtop ventilation may be necessary. 60/40 solder contains lead and the smoke from soldering should not be inhaled as it is toxic, gently blow the smoke away from your face. Lead-free solder is available.
5. Only hold the soldering iron by the handle, all other parts will be extremely hot, approximately 750° F
6. Hold wires or other components to be heated with insulated tweezers or clamps.
7. Always place the soldering iron back in the stand, never pass it off to another person.
8. Do not blow on the solder to cool it down, it is molten metal and could spray, instead wait a few seconds before soldering other components.
9. Always complete work over a table/bench so molten solder cannot drip onto you.
10. Keep cleaning sponge wet during use.
11. Always turn soldering iron off or unplug when not in use.
12. Always wash hands with soap and water after soldering.
13. Keep the soldering station free of electrical cables to prevent damage from the heated tip.
14. Know where the fire extinguishers are and how to operate them in case of emergency.
15. Seek medical attention if burned.
16. Used solder sponges and other contaminated waste must be disposed of as hazardous waste.

COMPLETION QUESTIONS

1. By _______________ the soldering iron tip often it will allow the solder to transfer from the soldering iron to the material easily.

2. To see if the soldering iron is hot test the tinning process by touching ________________ to the iron tip.

3. Only solder in areas with proper ________________.

4. Hold wires or other components to be heated with ________________ tweezers or clamps.

5. Do not connect an ________________ ________________ to materials being soldered.

6. Wipe excess solder off of the tip using a wet ________________ or a brass wire tip cleaner.

7. 60/40 solder contains ________________ and the smoke from soldering should not be inhaled.

8. Allow the solder to flow ________________ and ________________ the connection before removing the iron.

9. Always complete work over a ________________ so molten solder cannot drip onto you.

10. Always ________________ your hands after soldering.

WOOD BURNER

PART IDENTIFICATION

Identify the circled parts on the wood burner illustrated below.

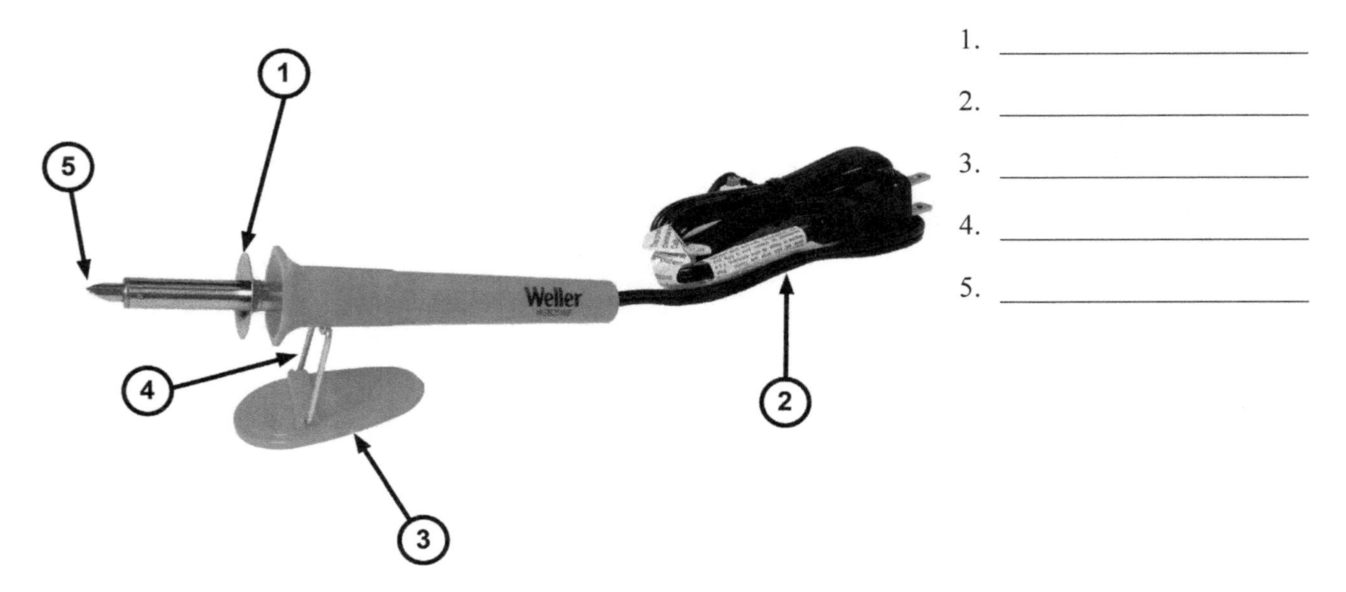

1. ____________________
2. ____________________
3. ____________________
4. ____________________
5. ____________________

SAFE OPERATIONAL PROCEDURES

1. Before you start, prepare all the tools necessary to complete the woodburning process.
2. Choose softer woods as they are easier to burn than harder woods.
3. Cut stock to get the desired shape and sand the wood surface smooth.
4. Transfer the design onto the wooden board.
5. Plug in the wood burner and wait for it to heat up.
6. Once the wood burner is at temperature use a scrap piece to test it. If it burns the wood instantly, the pen is hot enough to burn with.
7. Use a stand or docking station to keep the hot wood burning tip off the work surface when it is not in use.
8. If the tip comes loose while wood burning, use a pair of pliers to tighten it back up. Do not wood burn with a loose tip.
9. If you want to switch tips when in the middle of burning, let the wood burner cool down first and then replace the tip.
10. Keep a damp paper towel on hand to wipe off the tip every once in a while, as residue from the wood builds up.

11. Once complete, unplug the wood burner and let it cool before wrapping up the cord and putting it away.

GENERAL SAFETY PRACTICES

1. Wear approved eye protection. Tie up loose hair and remove loose jewelry.
2. Do not operate the wood burner without the instructor's permission, or without instructor supervision.
3. Make sure the wood you choose is dried, sanded smooth, and not chemically treated with stains or other finishes.
4. Do not leave a hot tool unattended; always turn off AND unplug the wood-burning tool before leaving the room.
5. Check the toxicity of the wood you will be using prior to burning a new piece.
6. Work on a clean, hard surface in a well-ventilated area; fans or fume extractors are always a good idea.
7. Avoid smoke inhalation and eye irritation by working with your face tilted to the side of the wood being burned.
8. Use a set of pliers and a ceramic dish for hot tips. This will allow you to safely switch, remove, and temporarily discard hot tips.
9. Be careful not to touch the hot tip to your skin, flammable objects like paper or the electrical cord.
10. Do not burn holes in the wood. Wood burning is for surface burning only.

COMPLETION QUESTIONS

1. Do not leave a hot tool ________________; always turn off AND unplug the wood-burning tool before leaving the room.
2. Before you start, prepare all the ________________ necessary to complete the woodburning process.
3. Use a set of ________________ and a ________________ dish for hot tips. This will allow you to safely switch, remove, and temporarily discard hot tips.
4. Make sure the wood you choose is dried, ________________ ________________, and not chemically treated with stains or other finishes.
5. Use a ________________ or ________________ ________________ to keep the hot wood burning tip off the work surface when it is not in use.
6. Work on a clean, hard surface in a ________________ - ________________ ________________; fans or fume extractors are always a good idea.
7. Avoid ________________ ________________ and eye irritation by working with your face tilted to the side of the wood being burned.

8. Once complete, ________________ the wood burner and let it cool before wrapping up the cord and putting it away.
9. Be careful not to touch the hot tip to ________________ - ________________ like paper or the electrical cord.
10. Once the pen is at temperature use a ________________ piece to test it.

3D PRINTER

PART IDENTIFICATION

Identify the circled parts on the 3D printer illustrated below.

1. ____________________
2. ____________________
3. ____________________

SAFE OPERATIONAL PROCEDURES

1. Before printing:
   a. Prepare all files to be 3D printed:
      (1) Be sure they are all sitting on the build plate and not above the build plate in the file set up software, check the scale of the parts to be printed
   b. Turn on the printer and let it run through the initializing process
      (1) Check that the build plate is clear, replace the build plate making sure it is level
      (2) Check that there is enough filament for the job to be run
   c. Send the file to be printed to the printer
2. During printing:
   a. Ensure that the filament is adhering to the build plate and laying nice and even rows of material, if the material is not adhering to the build plate stop the job; check the temperature and level the plate to avoid jamming the nozzle with filament. Clear the plate and restart the print.
   b. If the machine is equipped with doors do not open the doors during printing.
3. After the build is complete:

a. Remove all material from the build plate with a putty knife

b. If using a dissolvable filament for support place the printed parts in the dissolvable solution tank, wear goggles, gloves, and use a pair of tongs so as not to get the chemical on your skin. Always place parts inside the basket in the dissolvable solution tank so you can remove all of the parts when the material has dissolved.

4. Frequently check and level the build plate if the printer allows this setting

GENERAL SAFETY PRACTICES

1. Wear approved eye protection, hearing protection, and proper clothing. Tie up loose hair and remove loose jewelry.
2. Do not operate the machine without the instructor's permission, or without instructor supervision.
3. Always have a plastic putty knife nearby for removing parts from build plates that may be hot, do not touch the build plate, if it has a heating element it will be hot.
4. Do not touch the printhead/nozzle it is over 400 degrees hot.
5. Use a diagonal cutter to cut filament that has been removed from the printer before loading it to avoid jamming the extruder head with excess material.
6. Do not use material other than what the manual states, the machine may not be designed to work with various types of material, which could jam the nozzle, extruder, or potentially void the machines warranty.
7. Keep a record of all prints; the record may include whose parts are printed, the material being used, the run time of the print and so on. It is good to have this information for diagnosing problems and keeping track of printer use.

COMPLETION QUESTIONS

1. Always wear ____________________ ____________________ when cutting filament, removing parts from the build plate, or putting parts into or removing them from the dissolvable filament solution.
2. If the printer has a tank for dissolvable filament solution make sure you are wearing goggles, ____________________, and using tongs.
3. If the printer has a tank for dissolvable filament solution always place parts in the ____________________ so you can easily remove them from the solution when the support material has dissolved.
4. The ____________________ must be present while the printer is in operation.
5. Use a ____________________ ____________________ when removing parts from the build plate.
6. Ensure the build plate is clear and ____________________ before starting the job.
7. Do not open the ____________________ while the machine is in operation.

8. Make sure all parts are placed on the build ____________________ and not above it.
9. Check the ____________________ of the part to ensure they will fit the printable area.
10. Do not touch the ____________________ which operates above 400 degrees.

AUTOMOTIVE HOIST

PART IDENTIFICATION

Identify the circled parts on the automotive hoist illustrated below.

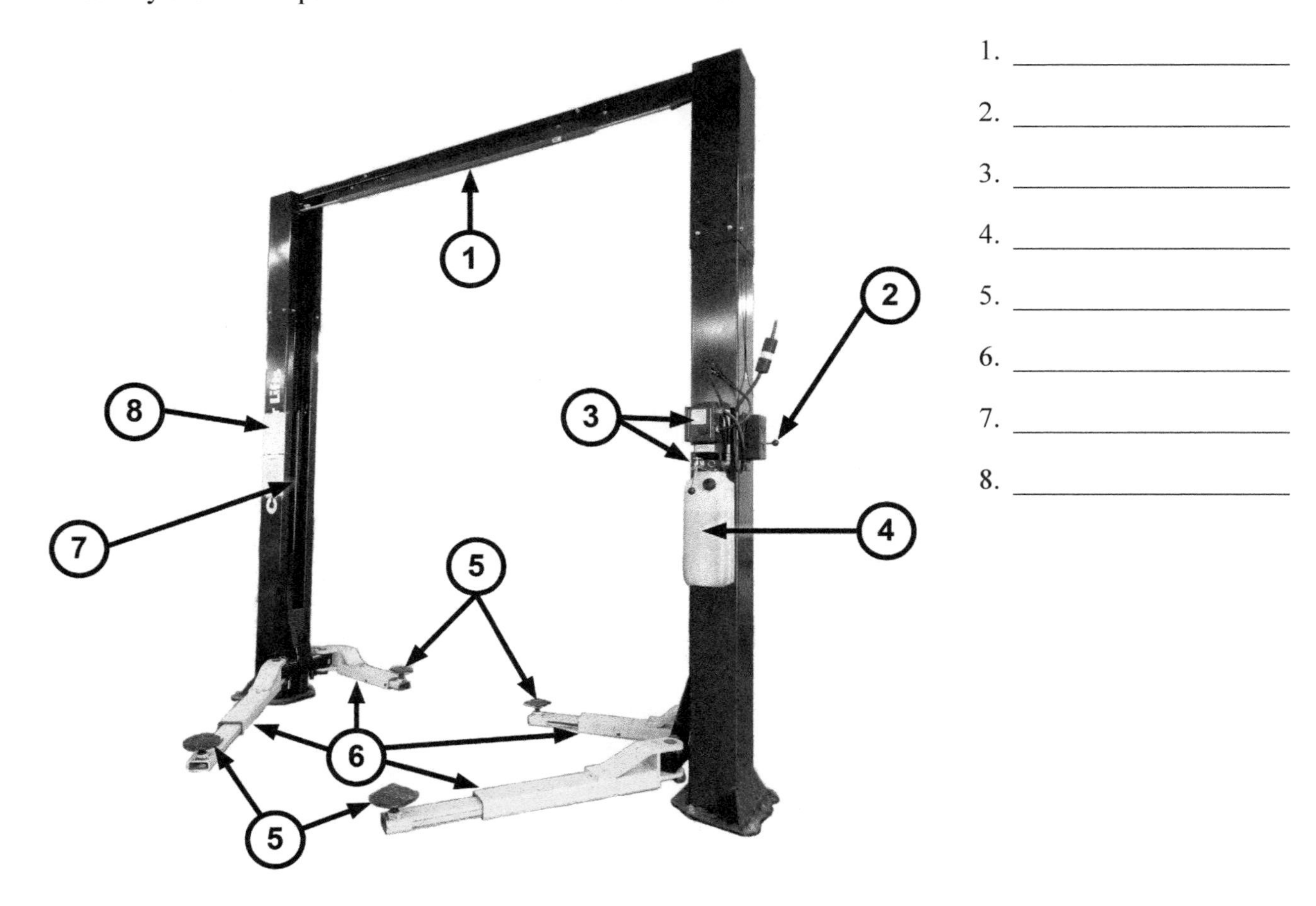

1. ____________________
2. ____________________
3. ____________________
4. ____________________
5. ____________________
6. ____________________
7. ____________________
8. ____________________

SAFE OPERATIONAL PROCEDURES

1. Lifting a vehicle on a four-post hoist.
   a. Remove vehicle cargo that could offset the vehicles center of gravity.
   b. Ensure the weight of the vehicle (GVWR can be found on the door pillar) does not exceed the lifting capacity (can be found by the controls) of the hoist.
   c. Ensure the hoist inspection card is current.
   d. Drive the vehicle centered between the posts.
   e. Lower the window, turn the vehicle off and place the keys in a secure location.
   f. Engage the parking brake and place wheel chocks on each side of a wheel.
   g. Give a verbal signal that you will be lifting, and lift the vehicle to the desired height.
   h. Lower the hoist onto the locks. Never work on a vehicle with the lift under pressure.

2. Lifting a vehicle on two-post hoist.
    a. Remove vehicle cargo that could offset the vehicle's center of gravity.
    b. Ensure the weight of the vehicle (GVWR can be found on the door pillar) does not exceed the lifting capacity (can be found by the controls) of the hoist.
    c. Ensure the hoist inspection card is current.
    d. Drive the vehicle centered between the post.
    e. Center the vehicle front to back according to lift manufacture's specifications.
    f. Lower the window, turn the vehicle off and place the keys in a secure location.
    g. Locate the four lift points of the vehicle which can be found in the owner's manual.
    h. If needed, raise the tabs to prevent the lift arms from damaging the body panels. The tabs are flipped opposing each other.
    i. Raise the lift close to the lift points and check to make sure lift points align with lifting pads.
    j. Continue lifting the vehicle six (6) inches off the ground and rock the vehicle lightly to check for stability.
    k. Once stability is established, give a verbal signal that you will be lifting, and lift the vehicle to the desired height, ensuring the roof of the vehicle does not come in contact with the crossbar of the lift.
    l. Lower the hoist onto the locks. Never work on a vehicle with the lift under pressure.
    m. If you are completing extensive work or work that could change the center of gravity, install an under hoist stand to prevent the vehicle from tipping.
3. Lowering a Vehicle
    a. Ensure all people, tool boxes, tools and under hoist stands are removed from below the vehicle.
    b. Verbally announce you are lowering the vehicle.
    c. Raise the vehicle slightly so it is no longer putting pressure on the locks.
    d. Hold the lock release lever and lower the vehicle to the floor.

GENERAL SAFETY PRACTICES

1. Wear approved eye protection, hearing protection, and proper clothing. Tie up loose hair and remove loose jewelry.
2. In some cases, safety goggles, face shields and bump caps may be required.
3. Do not operate the hoist without the instructor's permission, or without instructor supervision.
4. Maintain a constant awareness of the many hazards involved with lifting vehicles.
5. Never allow unqualified persons to enter the area.
6. Check the lifting points and adapters for corrosion, damage, slick or oily surfaces.
7. Remove any item in the vehicle that could affect the normal center of gravity.

8. Ensure the locking devices (latches) are properly engaged.
9. Never lower a vehicle onto the under hoist support stands.
10. Never try to stabilize a falling vehicle. Get out of the way!
11. Never try to alter or repair a lift. Immediately report any problems to your instructor. Only trained professionals are qualified and authorized to repair or modify the equipment.
12. Always lift the entire vehicle using all the arms at the same time.
13. Avoid standing directly in front of or behind the vehicle while lifting. Stand behind the post if possible.
14. If a vehicle does fall off the hoist, do not continue lowering the vehicle. Contact a certified vehicle recovery company and have them remove the vehicle.
15. Never lift with an occupant in the vehicle.

COMPLETION QUESTIONS

1. Remove vehicle _______________ that could offset the vehicles center of gravity.
2. Ensure the _______________ _______________ _______________ is current.
3. Drive the vehicle _______________ between the posts.
4. When using a 4 post lift, engage the _______________ _______________ and chock the wheels.
5. Give a_______________ _______________ that you will be lifting, and lift the vehicle to the desired height.
6. Lower the lift onto the _______________. Never work on a vehicle with the lift under pressure.
7. On a 2 post lift, locate the _______________ lift points of the vehicle which can be found in the owner's manual.
8. If you are completing extensive work or work that could change the center of gravity, install an _______________ _______________ _______________ to prevent the vehicle from tipping.
9. Never lift with an _______________ in the vehicle.
10. Always lift the_______________ vehicle using all the arms at the same time.

BELT/DISC SANDER

PART IDENTIFICATION

Identify the numbered parts of the belt/disc sander illustrated below.

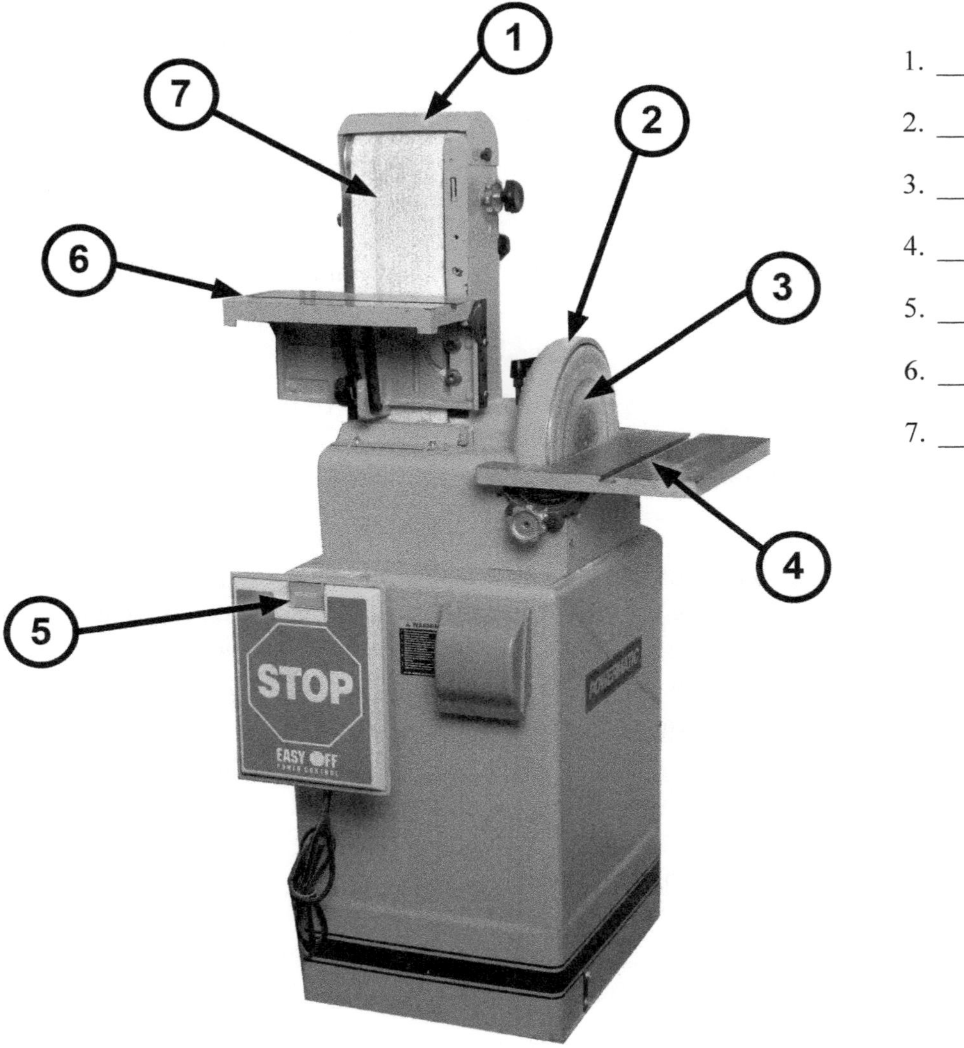

1. ____________________
2. ____________________
3. ____________________
4. ____________________
5. ____________________
6. ____________________
7. ____________________

SAFE OPERATIONAL PROCEDURES

1. Disc Sander
   a. The disc sander is useful for sanding or shaping flat and convex edges or end grain of material.
   b. The disc sander is sized by the diameter of the disc in inches.
   c. Select the coarseness of the abrasive disc according to the more common sanding jobs to be completed. Keep the sandpaper clean and in good condition.

d. The disc sander is usually equipped with an adjustable table on which a miter gauge is used to guide the material when sanding bevels and angles. A square should be used to check the angle between the table and the disc.

e. Adjust the table so the edge is 1/8" or less from the disc. Disconnect the machine before making adjustments.

f. Keep the disc guard in place at all times.

g. To sand or smooth the end of a board, place the board on the sander table so the disc sands in a downward direction on the board. DO NOT use the upward moving portion for sanding.

h. Turn the motor on, allow the motor to gain full speed, and then move the material carefully into the disc with light pressure.

i. Hold the material down against the table firmly at all times.

j. Move the material sideways (back and forth) slightly to reduce the heat caused by friction between the disc and the edge of the material. Remove the material from the disc often to give it an opportunity to cool.

k. If you are still experiencing burning, ensure the sandpaper is clean.

l. When the material is smooth or to the desired sanding line, remove it from the table and turn off the motor.

m. Do not leave the safety zone area until the disc has completely stopped.

2. Belt Sander

   a. The belt sander is useful for sanding or shaping flat and convex edges or end grain of material with the table. With the fence installed it can also do faces, and by removing the belt end guard and using the idler pulley, it can also be used to smooth concave contours.

   b. Select the coarseness of the abrasive belt according to the more common sanding jobs to be completed. Keep the sandpaper clean and in good condition.

   c. The belt sander is usually equipped with an adjustable table on which a miter gauge is used to guide the material when sanding bevels and angles. A square should be used to check the angle between the table and the belt.

   d. Adjust the table so the edge is 1/8" or less from the belt. Disconnect the machine before making adjustments.

   e. For concave contour sanding, you may remove the belt end guard and perform the sanding operation. Replace the guard after you are finished.

   f. For horizontal machines, make sure no one is standing in line with the belt.

   g. Avoid a kickback by working against the rotation of the belt.

   h. If using the fence, avoid getting fingers too close to the fence in order to prevent pinching.

   i. Turn the motor on, allow the motor to gain full speed, and then move the material carefully into the belt with enough pressure to keep the sandpaper cutting.

   j. Hold the material down against the table firmly at all times.

   k. Move the material sideways (back and forth) slightly to reduce the heat caused by friction between the belt and the edge of the material. Remove the material from the belt often to give it an opportunity to cool.

l. If you are still experiencing burning, ensure the sandpaper is clean.

m. When the material is smooth or to the desired sanding line, remove it from the table and turn off the motor.

n. Do not leave the safety zone area until the belt has stopped.

GENERAL SAFETY PRACTICES

1. Wear approved eye protection, hearing protection, and proper clothing. Tie up loose hair and remove loose jewelry.
2. Do not operate the machine without the instructor's permission, or without instructor supervision.
3. Never wear gloves while operating the sander.
4. With the machine off, inspect the sandpaper to make sure it is clean and free from damage. Never operate the machine if the sandpaper is loose, torn, or filled with sanding dust.
5. Make all adjustments with the motor off and the disc and/or belt completely stopped.
6. Allow the machine to reach full operating speed before beginning to sand.
7. Maintain a balanced stance and keep your body under control at all times. Do not overreach.
8. Do not allow hands or fingers to get near or touch the moving abrasive/sandpaper. On small or thin material, use a push stick or a jig to keep the hands from contacting the abrasive.
9. Be sure the table is locked in position before placing material on it and that its front edge is within 1/8" or less of the abrasive/sandpaper.
10. Do not overload the motor with excessive pressure; instead use light pressure and keep your workpiece moving to avoid burning.
11. The sanders are designed to smooth edges or end grain of material and not for removing excessive amounts from edges or ends of boards. Long periods of continuous sanding will overheat the disc and or belt, causing the material to be discolored and possibly damaging the machine. Cut to within ⅛" of your finished lined before using the sander.
12. If you are using a combination disc and belt sander, never allow two people to use both sanders at the same time.
13. Never clean the table with power on. Never use your hands to clear sawdust and debris; use a brush.
14. Before leaving the machine, the sander should be turned off and the abrasive disc, belt, should be completely stopped.
15. Keep the machine guards in place at all times when the machine is in use. Remove the energy source before removing guards to perform maintenance.

COMPLETION QUESTIONS

1. You can use the ______________________ sander or the ______________________ sander to sand flat areas and convex curves.

2. You can use the ______________________ sander with the end guard removed to sand concave shapes.

3. The ______________________ can be tilted for sanding bevel edges.

4. The disc sander can sand ______________________ of material while the ______________________ can sand edges and faces.

5. Sanding should be done on the side of the disc that is moving ______________________.

6. The edge of the tilting table should never be more than ______________________ inch from the disc.

7. The ______________________ ______________________ should be used when sanding mitered edges.

8. Holding your material in one spot or applying pressure will cause your material to ______________________.

9. A ______________________ could be used to check the angle between the table top and sanding disc.

10. Allow the sander to obtain full ______________________ before touching your material to the sandpaper.

BENCH/PEDESTAL GRINDER

PART IDENTIFICATION

Identify the numbered parts of the bench/pedestal grinder illustrated below.

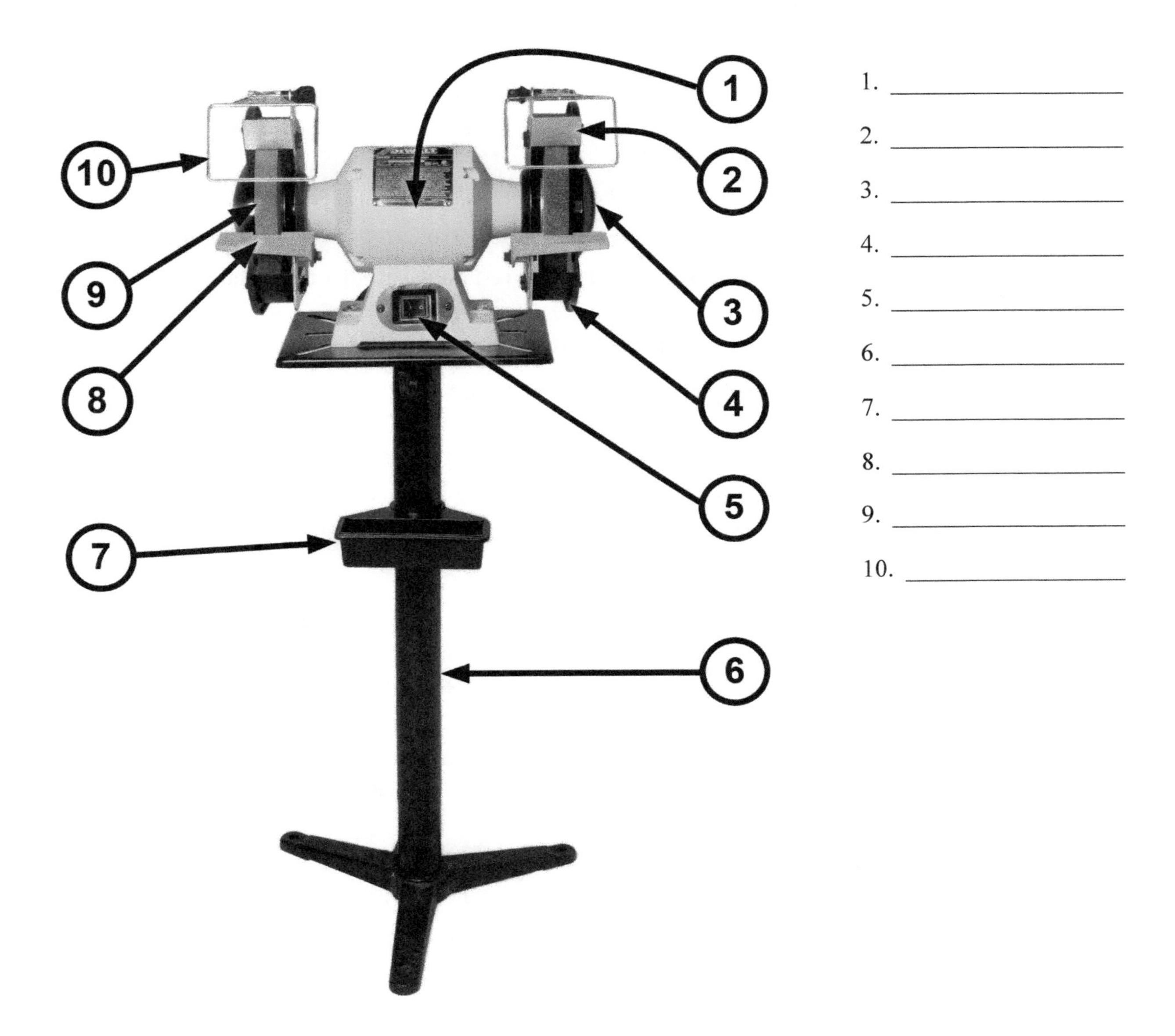

1. ____________________
2. ____________________
3. ____________________
4. ____________________
5. ____________________
6. ____________________
7. ____________________
8. ____________________
9. ____________________
10. ____________________

SAFE OPERATIONAL PROCEDURES

1. Tool sharpening:
   a. Selecting the grinding wheel:
      (1) When replacing a grinder wheel, be sure to "sound" the wheel before it is installed. Take a piece of string and place it through the arbor hole of the grinding wheel. Hold the grinding wheel up by the string and tap it lightly with a metal object (nail). A

ringing noise indicates the wheel is solid and usable; a dull sound indicates the wheel has an internal fracture and should be discarded.

(2) Select a 60-grit wheel for general tool sharpening.

(3) A soft grinding wheel is recommended for grinding hard materials.

(4) The grinding wheel should fit inside the housing.

(5) The arbor hole size should be .002" larger than the shaft size.

(6) The grinder speed should not exceed the speed listed on the side of the grinding wheel.

b. Maintaining the grinding wheel:

(1) Wear a dust mask when truing a grinding wheel. Never breathe the dust encountered when truing a grinding wheel.

(2) Use a wheel dresser to remove the glaze on the wheel after grinding soft material.

(3) Use a wheel dresser for straightening a grooved, rounded, or out-of-round wheel.

(4) Before dressing the wheel, adjust the tool rest on the grinder to a point even with the center of the wheel.

(5) Allow the grinder to reach full operating speed before the grinding operation is started.

(6) Do not stand in direct line with the grinding wheel when using the grinder. Stand to the side so that if a piece of the grinder wheel is thrown off, it will not hit you.

(7) Place the wheel dresser on the tool rest, gripping it with both hands.

(8) Move the wheel dresser back and forth across the wheel. Do not let the dresser pass off the edge of the wheel.

(9) Remove just enough material to clean and straighten the wheel.

(10) Use a combination square to make sure the face of the wheel is square with the side.

c. Sharpening a tool:

(1) Adjust the tool rest so that it is not more than 1/8 inch from the wheel and slightly above the center.

(2) Stand to one side and start the grinder. This will prevent the operator from being hit by wheel fragments if the grinding wheel should disintegrate when the grinder is turned on.

(3) Allow the machine to come up to speed before starting to grind.

(4) Carefully place the tool against the grinding wheel so its cutting edge is against the direction of rotation.

(5) Move the tool slowly across the face of the wheel.

(6) Do not allow the tool to overheat. Dip it in water or use a very light feed and stop to allow it to cool in the air.

GENERAL SAFETY PRACTICES

1. Wear approved eye protection, hearing protection, and proper clothing. Tie up loose hair and remove loose jewelry.

2. Do not operate the machine without the instructor's permission, or without instructor supervision.

3. Do not wear gloves which may get caught between the grinding wheel and tool rest when grinding.

4. Make sure to use ferrous metal only with stones made for ferrous material. Non-ferrous material will plug up and cause potential hazard.

5. Be sure the housing, wheel guards and safety shields are in place around the grinding wheel. They are there for protection in case the grinding wheel should break.

6. The spark deflector should be adjusted to within 1/16 inch of the grinding wheel.

7. Do not use a wheel that vibrates excessively. Dress such grinding wheels with the wheel dresser to make them turn true. If the wheel cannot be dressed, check the grinder for a bent arbor. Do not continue to use a grinder with a bent arbor.

8. Never use a grinding wheel that is cracked or broken.

9. Make sure the operator can see the work well by keeping the safety shields clean and work lights functional.

10. Wear a face guard even though the grinder has a glass shield.

11. Do not hold material being ground in such a way that fingers may contact the wheel surface.

12. Do not use the side of the wheel for rough grinding. This will place too much stress on the wheel and cause it to break.

13. Do not force material into the wheel and do not put excessive pressure on the grinding wheel. Excessive pressure may cause pieces of the wheel to break and be thrown at the operator.

14. Do not use the grinder if another person is within the grinder operating zone or in direct line of the grinding wheel action.

15. Never remove the paper from the sides of the grinding wheel.

16. Do not use a grinding wheel that is worn to less than 1/2 of its original diameter. For example, an 8" wheel should not be used at less than 4" in diameter.

17. Do not leave the grinder without turning it off and making sure it has completely stopped rotating.

18. Identify the location of the tool shut off for use in an emergency situation.

COMPLETION QUESTIONS

1. Soft stones are recommended for grinding ______________________ materials.

2. The arbor hole size should be ______________________ inch(es) larger than the shaft size.

3. A______________________ ______________________ is used to straighten a grooved, rounded, or out-of-round grinding wheel.

4. The wheel dresser should not pass off the ______________________ of the grinding wheel.
5. A ______________________ can be used to determine if the face of the wheel is square with its side.
6. Before sharpening a tool, adjust the tool rest so it is not more than ______________________ inch(es) from the wheel.
7. The ______________________ ______________________ of the tool being sharpened should be against the direction of rotation.
8. The tool should be moved slowly across the ______________________ of the wheel
9. The tool can be dipped in ______________________ to keep it from overheating.
10. Stand ______________________ ______________________ ______________________ when starting the grinder.

BENCH BUFFER

PART IDENTIFICATION

Identify the circled parts on the bench buffer illustrated below.

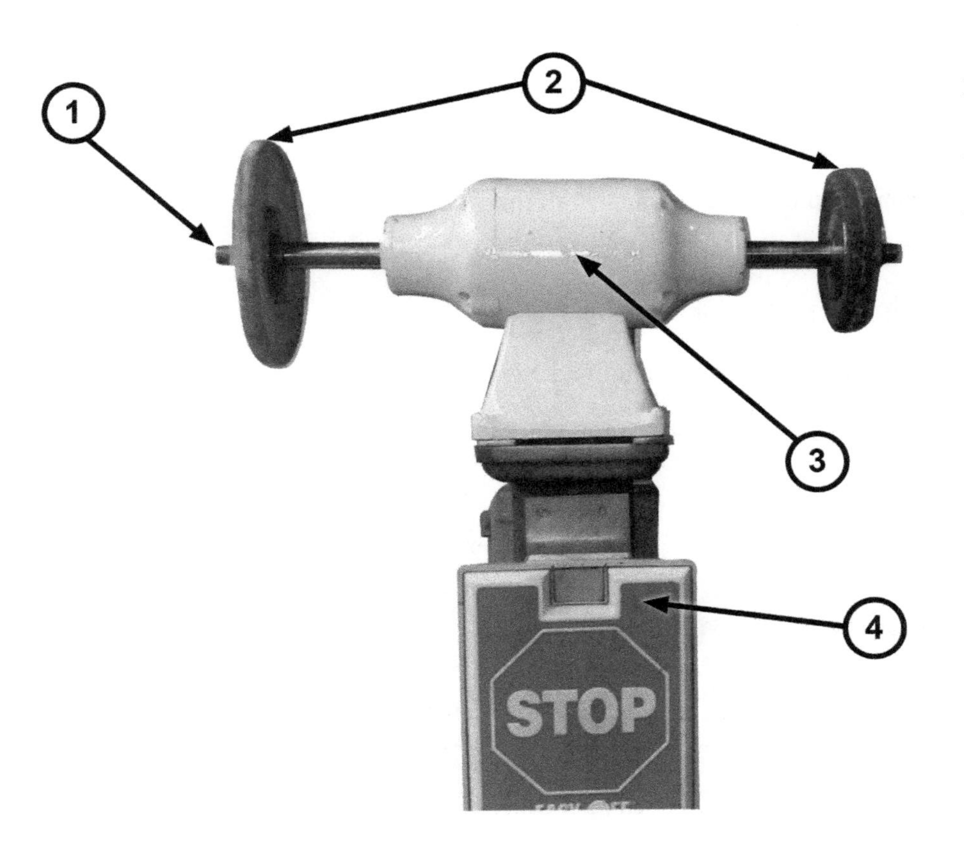

1. ____________________

2. ____________________

3. ____________________

4. ____________________

SAFE OPERATIONAL PROCEDURES

1. Review and understand information provided in the bench buffer operators manual.
2. Before using the bench buffer, inspect for damage or disrepair. Report faulty equipment to your instructor.
3. Be sure to have adequate light while working with the bench buffer machine.
4. Hold on to your work piece, it can become hard to grip or become warm.
5. In the event of an emergency, shut off the bench buffer immediately.
6. Turn off the bench buffer and disconnect the power source prior to making adjustments.
7. Always clean the bench buffer work area upon completion of your work. Never use compressed air to clean the buffing machine, use a chip brush and rag.

GENERAL SAFETY PRACTICES

1. Wear approved eye protection, hearing protection, and proper clothing. Tie up loose hair and remove loose jewelry.
2. Do not operate the machine without the instructor's permission, or without instructor supervision.
3. Wear a dust mask or have a dust collection system for the bench buffer.
4. Never wear gloves when working on moving machinery.
5. Check to be sure all workspaces and walkways are clear of slip/trip hazards.
6. Ensure both ends of the machine spindle are covered at all times.
7. Use the appropriate type of mop and polishing compound for the task.
8. Only use the front of the buffing wheel, do not work on the side of the mop.
9. Only one person may operate the buffing wheel at a time. Never leave the machine running and unattended.

COMPLETION QUESTIONS

1. Do not use a ______________ bench buffer, immediately report damaged equipment to your instructor.
2. Never leave the bench buffer ______________ and unattended.
3. Never use the bench buffer without getting the instructors ______________.
4. Eye protection must be worn at all times when ______________ the bench buffer.
5. Loose ______________ and ______________ can get caught in moving parts of the bench buffer.
6. Only use the ______________ of the buffing wheel, do not work on the side of the mop.
7. In the event of an emergency, ______________ ______________ the bench buffer immediately.
8. Never use ______________ ______________ to clean the buffing machine, use a chip brush and rag.
9. Be sure to have ______________ light while working with the buffing machine.
10. Never wear ______________ when working on moving machinery.

BRAKE LATHE

PART IDENTIFICATION

Identify the circled parts on the brake lathe illustrated below.

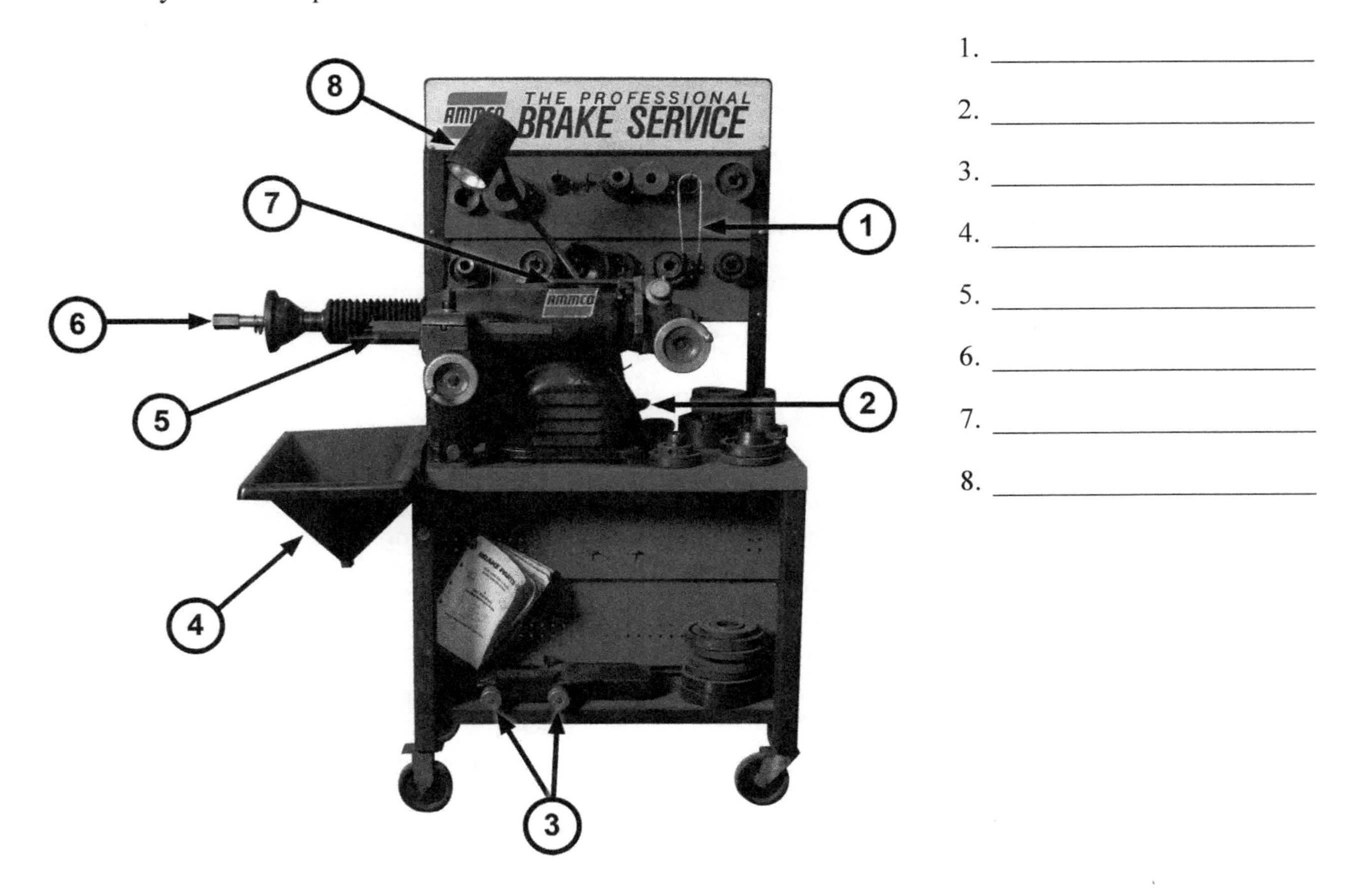

SAFE OPERATIONAL PROCEDURES

1. Inspect your brake lathe machine on a regular basis, look for moisture, dirt and rust.
2. Check to be sure all tools and accessories are clean and in good working order.
3. Position your body and keep hands & fingers away from the rotating brake areas.
4. Be cautious when the drum or rotor is rotating on the brake lathe machine.
5. Inspect the spindle carefully for wear and defects before spinning.
6. Be sure the drum or rotor are the correct size for the machine.
7. Be sure the safety shield is in place when machining a drum or rotor.
8. Use proper tools to install or detach the drum or rotor from the machine.

GENERAL SAFETY PRACTICES

1. Wear approved eye protection, hearing protection, and proper clothing. Tie up loose hair and remove loose jewelry.
2. Do not operate the machine without the instructor's permission, or without instructor supervision.
3. Safety toe boots are highly recommended.
4. Maintain a constant awareness of the many hazards involved with machining drums.
5. Never allow unqualified persons in the area while you are working.
6. Check the rotating spindle for corrosion, damage or sticky operation.
7. Ensure the locking devices (clamps) are properly working.
8. Never machine a drum or rotor past its recommended cutting speed.
9. Never try to repair the brake lathe machine, report any problems to your instructor.

COMPLETION QUESTIONS

1. Stand to the ________________ when a brake lathe machine is spinning.
2. ________________ on the drum or rotor will affect the end result when machining.
3. It's important to check the ________________ of the drum or rotor before machining.
4. Use brake wash, the ________________ and ________________ to clean a machined rotor.
5. Locking devices such as ________________, should be properly working.
6. Report all problems to your ________________, never try to repair the machine.
7. Always wear your ________________ ________________ ________________ when operating the brake lathe machine.
8. Never remove the ________________ on the brake lathe machine.
9. Use the correct cutting ________________ and operate them correctly.
10. Never allow ________________ persons in the area while you are working.

CNC LATHE

PART IDENTIFICATION

Identify the circled parts on the CNC lathe illustrated below.

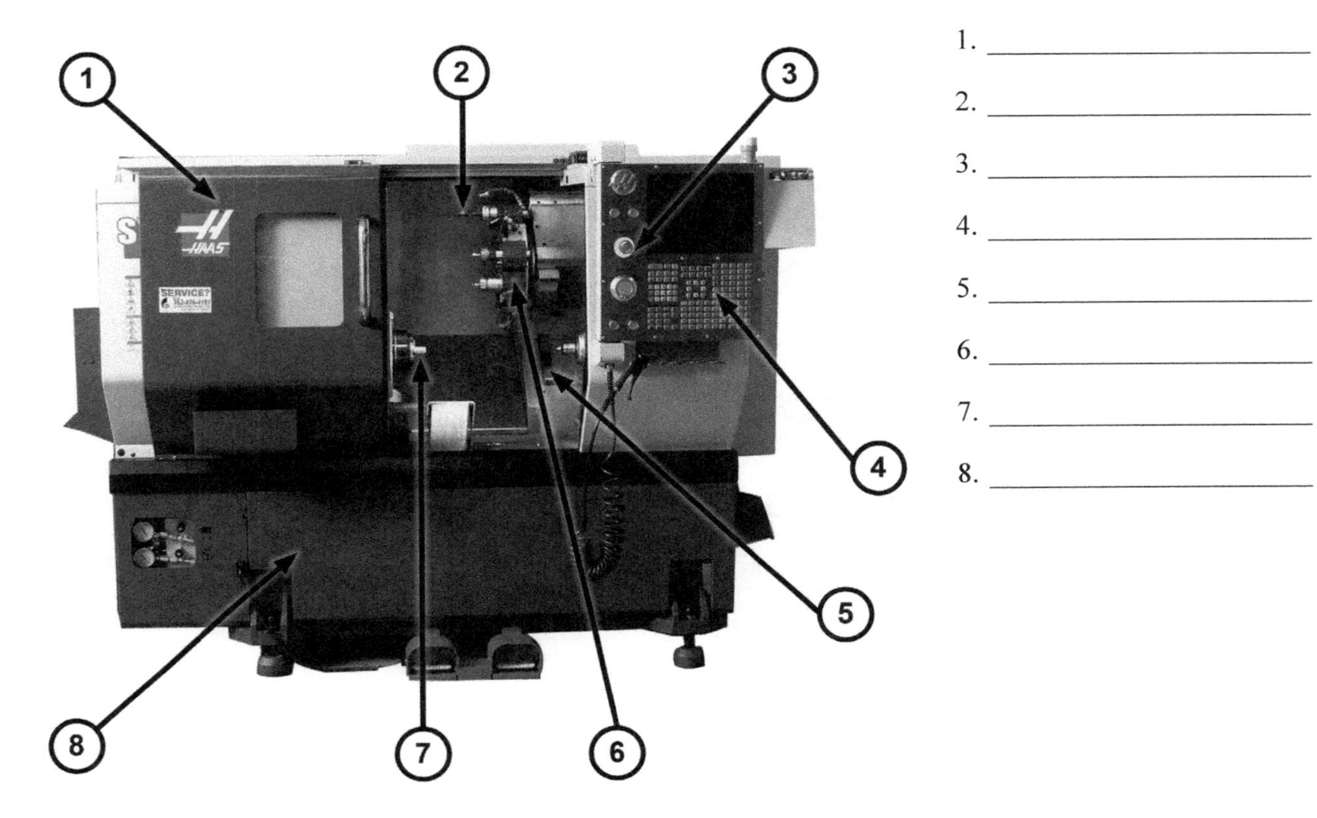

1. ________________________
2. ________________________
3. ________________________
4. ________________________
5. ________________________
6. ________________________
7. ________________________
8. ________________________

SAFE OPERATIONAL PROCEDURES

1. Check the machines air pressure.
2. Turn the control on and allow the machine to complete its start-up process.
3. Ensure the tool is away from the part and home both the X and the Z axis. This will be fully to the right and away from the operator.
4. Make sure the machine is clean and free of chips. If you have to remove chips, always wear gloves and safety glasses.
5. Check the coolant level and make sure the coolant reservoir is ¾ full.
6. Clean the chip catch tray above the coolant reservoir.
7. Engage the E-Stop.
8. Check the work holding device for damage and make sure it will hold your material appropriately.

9. Load the tools you will need for the operations you are completing.
10. Load your material in the work holding device.
11. Make a facing skim cut and set the Z-work coordinate.
12. Make a diameter skim cut, measure the diameter and enter the measurement in for the x-work coordinate.
13. Set all tool offsets of using multiple tools.
14. Load the correct program.
15. Press the cycle start button, keeping your hand on the feed stop button. You can also run the program in single line mode to slow the program down.
16. Once the program is successful and the machine has stopped, press the E-Stop button, remove the part being careful of the newly cut edges. They have burrs that are very sharp. Then you can clean the chip conveyor.
17. After all parts are completed, put all tools away and clean up any chips and spills around the machine.

GENERAL SAFETY PRACTICES

1. Wear approved eye protection, hearing protection, and proper clothing. Tie up loose hair and remove loose jewelry.
2. Do not operate the machine without the instructor's permission, or without instructor supervision.
3. CNC machines move automatically and will injure the operator if you are in the path of the tool or part.
4. Only one person should be operating the machine at a time.
5. Understand the operation of all the stop buttons on the machine. Use the E-Stop button if you are unsure or hear an incorrect sound coming from the machine.
6. The E-Stop is the only way to guarantee the machine will not move automatically, you should keep your hand over the E-stop at all times during operation so you can quickly stop the machine if needed.
7. If equipped with door interlocks, always keep the doors closed when the machine is operating and never attempt to override the system.
8. Always clean up spills immediately because coolant is slippery. Wear appropriate Personal Protective Equipment to handle the coolant.
9. The electrical cabinet and air/lube cabinet have high power in them and should only be serviced by authorized personnel.
10. Keep clear of the chip conveyor while it is in operation.
11. Chips can be razor sharp so any handling of them should be done with leather gloves on.

12. Cutting tools are sharp. Refrain from touching the cutting edges.
13. When cleaning parts with compressed air, never aim the air nozzle at yourself or another person. Air pressure can damage the body.
14. Remove work gloves and loose clothing when the machine is in operation. Loose articles are dangerous around the spinning material.
15. Read all SDS sheets with the materials you will be handing to understand how to protect yourself.
16. Keep your work area clean. Put away tools when you are completed with your task.

COMPLETION QUESTIONS

1. The _______________ is the only way to guarantee the machine will not move automatically.
2. Check the coolant level and make sure the coolant reservoir is _______________ full.
3. If equipped with door interlocks, always keep the doors _______________ when the machine is operating and never attempt to override the system.
4. Check the work holding device for _______________ and make sure it will hold your material appropriately.
5. After all parts are completed, put all _______________ away and clean up any chips and spills around the machine.
6. CNC machines move _______________ and will injure the operator if you are in the path of the tool or part.
7. Keep clear of the chip _______________ while it is in operation.
8. Chips can be razor sharp so any handling of them should be done with leather _______________ on.
9. Remove work gloves and loose _______________ when the machine is in operation. Loose articles are dangerous around the spinning material.
10. When cleaning parts with compressed air, never aim the air nozzle at yourself or another person. Air _______________ can damage the body.

CNC MILL

PART IDENTIFICATION

Identify the circled parts on the CNC mill illustrated below.

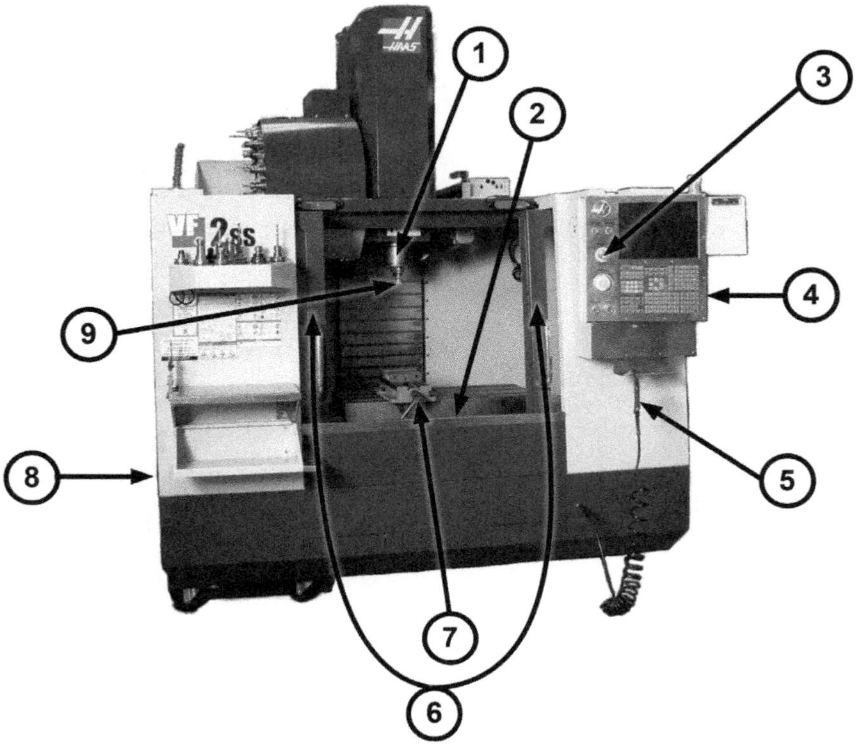

1. ____________________
2. ____________________
3. ____________________
4. ____________________
5. ____________________
6. ____________________
7. ____________________
8. ____________________
9. ____________________

SAFE OPERATIONAL PROCEDURES

1. Start with a clean machine. Clean any chips from the ways and the inside of the cabinet before you start working.
2. Load the tool(s) you will be using into the machine.
3. Load the material and securely clamp it into the work holding device.
4. Ensure air pressure is at the correct level coming into the machine.
5. Check the coolant tank. It should be at least ¾ full.
6. Power the machine on, disengage the E-Stop and home the machine.
7. Load the correct G-Code Program into the machine or enter conversational programming mode.
8. Set the work offset on the machine. The work offset tells the machine where your part is.

9. Set the tool offset(s) on the machine. The tool offset tells the machine how far the tool is below the spindle.

10. Setting the work offset so the machine will cut off your part, test the program. Have your hand on the E-Stop Button in case the machine looks like it will crash. Utilize single block mode or the cycle start/feed stop button to perform more precise movements.

11. Run the program with the correct work offset in place to cut the part, listening and watching for any problems in the program. Immediately hit the E-Stop if there is a problem.

12. Any time the program is running, keep all body parts outside of the footprint of the machine. It moves rapidly in all directions.

13. When program is complete press the E-Stop button to put the machine in a safe state.

14. Open the doors, unclamp the material and remove the chips and coolant with compressed air.

15. Remove tools from the machine and return them to their storage locations.

16. Clean the machine of all chips and coolant before leaving the area.

GENERAL SAFETY PRACTICES

1. Wear approved eye protection, hearing protection, and proper clothing. Tie up loose hair and remove loose jewelry.

2. Do not operate the machine without the instructor's permission, or without instructor supervision.

3. Give your full attention to the machine while it is running.

4. Only allow one operator on the machine at a time. Never have someone running the control panel while another person is working near the spindle.

5. Tools and cutters are razor sharp, never handle the sharp end.

6. If there is a problem, press the E-Stop button immediately. It will stop all movement including the spindle.

7. Any time you are working in the machine's envelope the E-Stop should be pressed.

8. The electrical control cabinet and the compressed air cabinet have high power levels and should only be opened by an authorized person.

9. If equipped with a chip auger, never clean the auger without the E-Stop button pressed.

10. Chips and shavings can be razor sharp. Always use brushes and compressed air to clean up.

11. Never use the compressed air on your body or clothing to blow off chips. Compressed air has enough pressure to cause injury.

12. Freshly cut parts have sharp burrs. Carefully deburr parts after they come out of the machine.

13. Always keep your work area clean. Clean up any coolant spills immediately.

COMPLETION QUESTIONS

1. Clean any chips from the ________________ and the inside of the cabinet before you start working.
2. Check the coolant tank. It should be at least ________________ full.
3. The ________________ offset tells the machine where your part is.
4. The ________________ offset tells the machine how far the tool is below the spindle.
5. When the program is complete, press the ________________ button to put the machine in a safe state.
6. Only allow ________________ operator on the machine at a time.
7. If there is a problem, press the E-Stop button immediately. It will stop all movement including the ________________.
8. The electrical control cabinet and the compressed air cabinet have high power levels and should only be opened by an ________________ person.
9. Chips and shavings can be razor sharp. Always use ________________ and compressed air to clean up.
10. Never use the ________________ ________________ on your body or clothing to blow off chips.

CNC PLASMA CUTTER

PART IDENTIFICATION

Identify the circled parts on the CNC plasma cutter illustrated below.

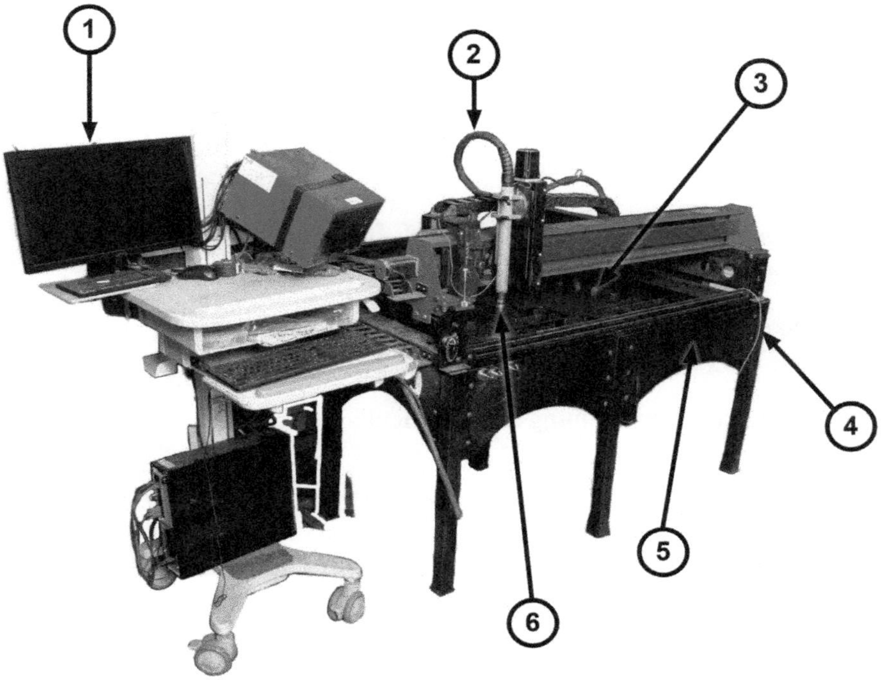

1. ______________________
2. ______________________
3. ______________________
4. ______________________
5. ______________________
6. ______________________

SAFE OPERATIONAL PROCEDURES

1. Make sure that the material is securely clamped to the corners of the table and grounded.
2. Select the type of cutting nozzle, air pressure, cutting speed, piercing time and adjust the current of the CNC plasma power supply according to the thickness of the material.
3. Any time the program is running, keep all body parts outside of the footprint of the machine. It moves rapidly in all directions.
4. Set the work offset so the machine will cut off your part and then test the program. Have your hand on the E-Stop Button in case the machine looks like it will crash.
5. Run the program with the correct work offset in place to cut the part, watching and listening for any problems in the program. Hit the E-Stop if there is a problem.
6. When program is complete press the E-Stop button, then turn off all power and air sources.
7. After the work is completed, properly store ground cable, clean the machine of all chips, slag, and coolant before leaving the area.

GENERAL SAFETY PRACTICES

1. Wear approved eye protection, hearing protection, and proper clothing. Tie up loose hair and remove loose jewelry.
2. Do not operate the machine without the instructor's permission, or without instructor supervision.
3. Do not allow any unauthorized personnel to use CNC plasma cutting equipment without supervision or permission.
4. Wear a minimum of a #5 shaded lens for the plasma arc cutter process.
5. Never operate the CNC plasma cutter when wearing articles of clothing that are made of synthetic materials.
6. Never place any part of your body in direct contact with areas of the machine that can pinch and crush when powered on.
7. Keep area clear of bystanders, make sure that no one but the person operating the machine has access to the computer controls.
8. Maintain a safe distance between your work and cutting area, particularly if items in your workspace are flammable.
9. Keep a fire extinguisher accessible at all times.
10. After cutting, use gloves, pliers or tongs to remove your material from the cutting table.
11. Never use the CNC plasma cutter in areas where combustible or explosive gases or materials are located.
12. Insulate yourself from material and ground, wear dry gloves and clothing.
13. When not cutting or gouging, make certain no part of the electrode circuit is touching the material or ground. Accidental contact can cause overheating and create a fire hazard.
14. Plasma cutting or gouging may produce hazardous fumes and gases. Avoid breathing these fumes and gases. Have adequate ventilation keep fumes and gases away from the work zone.
15. Always be sure the work cable makes a good electrical connection with the material being cut or gouged. The connection should be as close as possible to the area being cut or gouged.
16. Disconnect from the power source before working on the equipment.

COMPLETION QUESTIONS

1. The minimum shaded lens that should be used for plasma arc cutting is ________________.
2. You should always wear ________________ clothes when operating the CNC plasma cutter.
3. Keep area clear of bystanders, make sure that no one but the person operating the machine has ________________ to the computer controls.

4. When program is complete press the ________________ button, then turn off all power and air sources.
5. Never place any part of your body in direct contact with areas of the machine that can ________________ and ________________ when powered on.
6. Keep a ________________ accessible at all times
7. The plasma cutting process produces lots of ________________ fumes and therefore must be well ventilated.
8. Insulate yourself from material and ground, wear ________________ gloves and clothing.
9. ________________ from the power source before working on the equipment.
10. Securely clamp the material to the table and make sure it is ________________.

CNC ROUTER

PART IDENTIFICATION

Identify the circled parts on the CNC router illustrated below.

1. ____________________
2. ____________________
3. ____________________
4. ____________________
5. ____________________

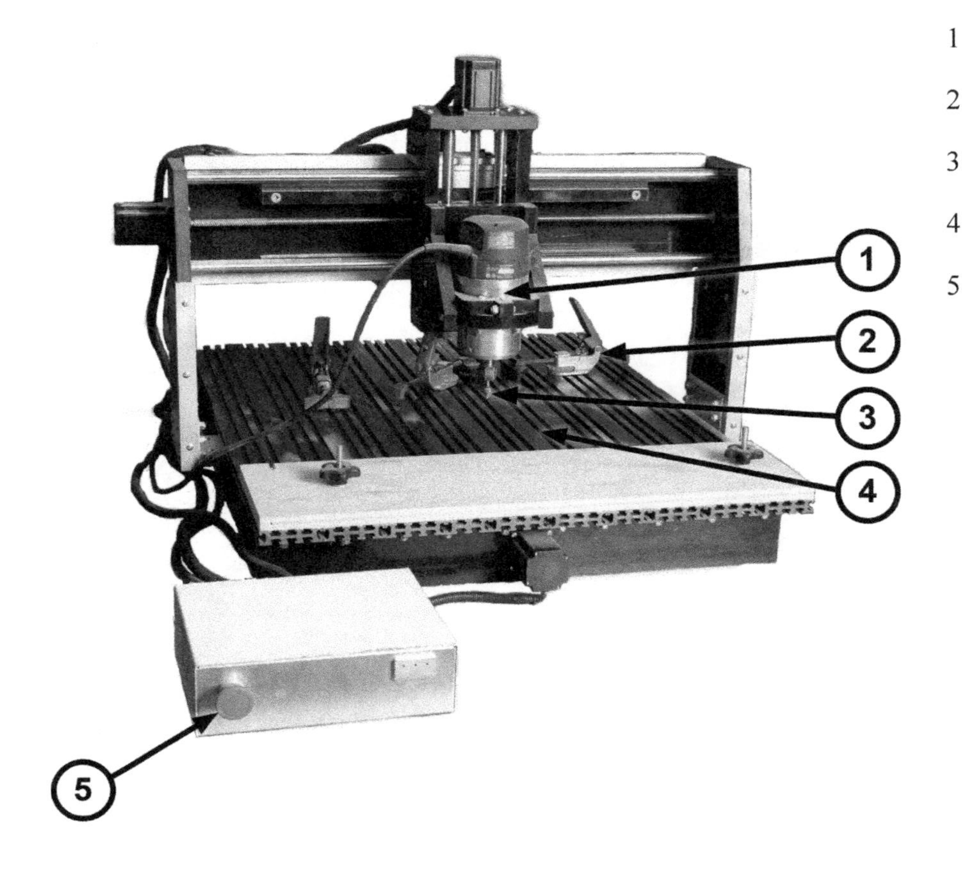

SAFE OPERATIONAL PROCEDURES

1. Ensure you are familiar with CNC ‘nesting’ and ‘tool-pathing’ software functionality.
2. Locate & ensure you are familiar with the operation of the ON/OFF and emergency stop controls.
3. Ensure that the guard door and safety devices are in position and secured.
4. Only machine materials that are suitable for this routing process.
5. Ensure that the router cutting bit size conforms to specifications. The machine must be isolated while any adjustments are made to the cutter head of tool array.
6. Ensure all cutters are sharp and free of resin build-up or wear.
7. Adjust the waste collector shroud and coolant system correctly for maximum efficiency.

8. Be aware of any other personnel in the immediate vicinity and ensure the area is clear before using this equipment.
9. Familiarise yourself with all electrical and mechanical operations and controls, including any handheld keypad interface remote control.
10. Never attempt to program this CNC machine without proper training.
11. Never pre-program any CNC router to perform operations beyond the capacity of the machine.
12. Confirm all CNC programming instructions for the router.
13. Ensure the work piece is securely held flat to the base of the machine.
14. Ensure that the tool bit array tracking remains unobstructed during the routing operation.

GENERAL SAFETY PRACTICES

1. Wear approved eye protection, hearing protection, and proper clothing. Tie up loose hair and remove loose jewelry.
2. Do not operate the machine without the instructor's permission, or without instructor supervision.
3. Do not operate a CNC router in a damp or moist environment.
4. Never leave the room while your machine is routing.
5. Do not put any part of your body inside the frame of the machine when it is in operation.
6. Always make sure you have inserted the correct bit for the job you are doing.
7. Check that the horsepower is suitable for the material being cut.
8. Make sure that all workpieces are securely clamped down.
9. Do not force material into the router.
10. Don't try to increase the depth of the cut by inserting less of the shank into the collet.
11. If the bit is acting up or breaks, shut down the machine immediately.
12. Do not use non-routing tools, even though their shanks may fit in the collet.
13. Clean the machine and the floor around it after every use.
14. Empty the vacuum bag frequently, or clear sawdust as needed.
15. Maintain your CNC router, replace worn parts as and when needed.
16. Clean the floor and work area around the machine.
17. Inspect the equipment to ensure no obvious defects: damaged chuck, dull or cracked tools, damaged shields.

COMPLETION QUESTIONS

1. Never operate any CNC machine without prior ____________________.
2. Always ____________________ the RPM and feed rate settings before machining.
3. Never ____________________ a CNC machine while it is running.
4. Make sure you are familiar with procedure to stop the CNC machine in case of an ____________________ situation.
5. Never ____________________ CNC machine if you are not feeling 100% and cannot concentrate.
6. Make sure that the correct____________________ is loaded and never rush a program execution on CNC machines.
7. Never override any ____________________ safety functions on the CNC machines.
8. Remember to follow all necessary ____________________ safety precautions before operating a CNC machine.
9. Make sure the workpiece is ____________________ securely.
10. When installing tools, do not forget to set each individual ____________________.

DRILL PRESS

PART IDENTIFICATION

Identify the numbered parts of the drill press illustrated below.

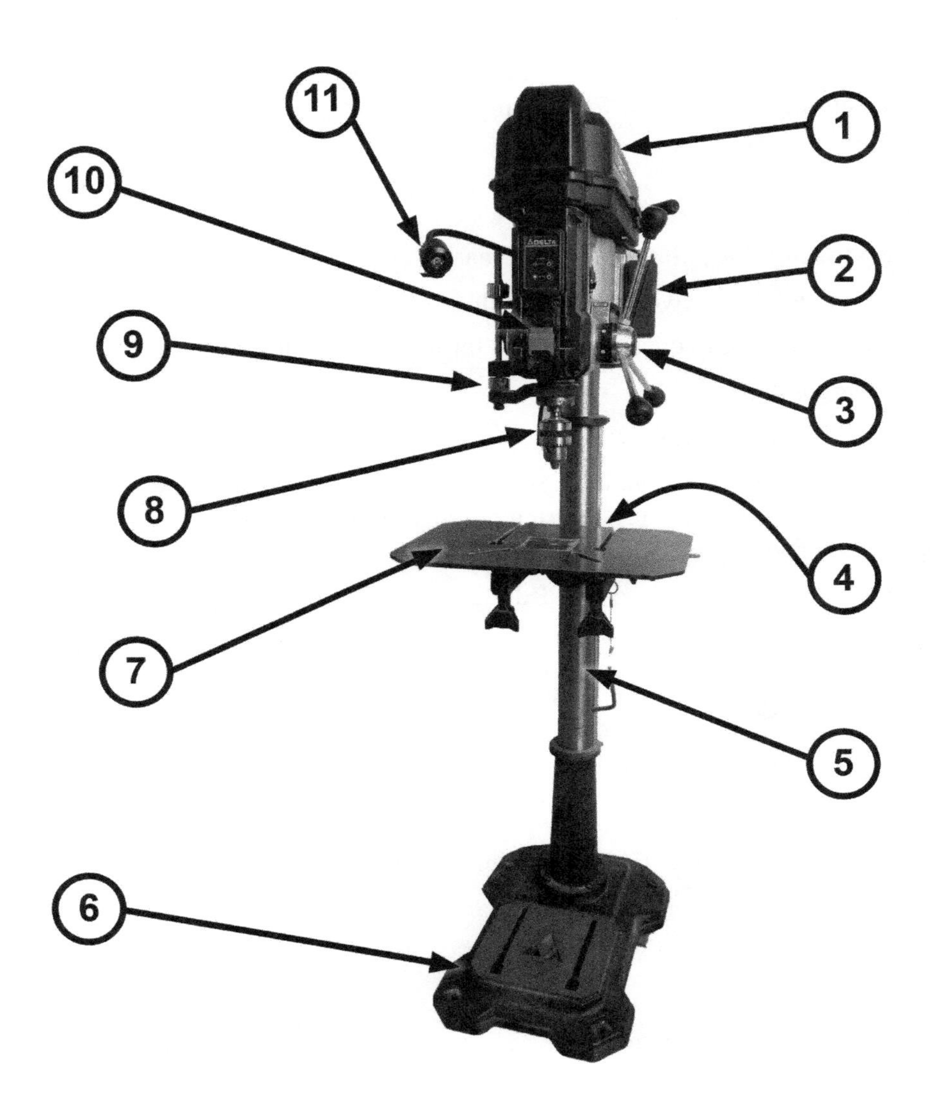

1. ____________________
2. ____________________
3. ____________________
4. ____________________
5. ____________________
6. ____________________
7. ____________________
8. ____________________
9. ____________________
10. ____________________
11. ____________________

SAFE OPERATIONAL PROCEDURES

1. General drilling:
   a. Use a drill gauge to determine the drill bit size.
   b. Be sure the drill bit is sharp and free of defects.
   c. Make sure the material is secure on the table.
   d. Feed the drill bit into the material at a slow and constant rate.
   e. When drilling round material, use a "V" block vise to prevent material from spinning.

f. Place the material on a block of wood to prevent damage to the drill press table.

2. Wood drilling:
   a. Using a combination square, draw two lines perpendicular to each other with their intersection being the center of the hole to be drilled.
   b. Mark the center of the hole to be drilled
   c. Select proper drill speed according to the operator manual.
   d. When drilling into wood: plunge your drill bit up and then back down periodically to allow it to cool in between cutting and to allow for clearing of the chips.

3. Metal drilling:
   a. Clamp the material to be drilled securely to the table.
   b. Use a scribe or scratch awl to mark the point for a punch mark.
   c. Center punch a mark large enough to receive the point of the drill bit.
   d. Select the proper drill speed according to operator manual for the material and size.
   e. Drill a pilot hole for holes larger than ½".
   f. Use cutting oil when drilling into hard metal.
   g. When drilling into metal: If you are getting small chips, you are either feeding too slow or your drill bit is dull. You should get nice curls coming out of your material when you are feeding at the proper speed and your drill bit is sharp.
   h. Clamp sheet metal between two blocks of wood and drill through wood and metal to prevent it from bending.

GENERAL SAFETY PRACTICES

1. Wear approved eye protection, hearing protection, and proper clothing. Tie up loose hair and remove loose jewelry.
2. Do not operate the machine without the instructor's permission, or without instructor supervision.
3. Place the long end of the material to the left so it will hit the post/column and not the operator should the material slip and start rotating.
4. Be sure chuck is tight and the drill bit is straight and centered in the chuck. The chuck should always be tightened in all three tightening positions.
5. Always remove the chuck key from the chuck immediately after using it and place it back in its proper place.
6. Do not wear gloves when operating the drill press. They may get caught in the rotating bit.
7. Do not feed the drill bit into the material faster than it can be easily cut. Also, do not feed too slowly or you will burn the bit.
8. As a rule, larger drill bits run slower, while smaller drill bits run faster.
9. Remove the chips/debris from the machine with a brush.

10. Do not reach around the machine.
11. Do not talk to anyone while operating the machine.
12. Do not drill into a container that may have contained gasoline or other flammable materials.
13. Support long material being drilled.
14. To avoid chip out on the back side, slow the drill feed when it is breaking through the material to finish the hole.
15. Hold cylindrical material in a “V” block to prevent it from spinning.

COMPLETION QUESTIONS

1. A drill gauge can be used to determine ______________________ ______________________ ______________________.
2. A ______________________ block should be used to hold cylindrical material.
3. A ______________________ ______________________ placed under the material being drilled will prevent damage to the table.
4. When drilling into wood: Plunge the __________________ up and then back down periodically.
5. A __________________ punch should be used before a hole is drilled into a piece of metal.
6. As a rule, larger drill bits run __________________, while smaller drill bits run __________________.
7. When drilling a hole in metal larger than 1/2”, drill a __________________ __________________.
8. __________________ oil should be used when drilling into hard metal.
9. When drilling into metal: If you are getting small chips, you are either feeding too slow or your drill bit is dull. You should get nice __________________ coming out of your material when you are feeding at the proper speed and your drill bit is sharp.
10. The long end of a piece of material should be kept to the __________________ to prevent injury to the operator should the material start to rotate.

DRUM SANDER

PART IDENTIFICATION

Identify the circled parts on the drum sander illustrated below.

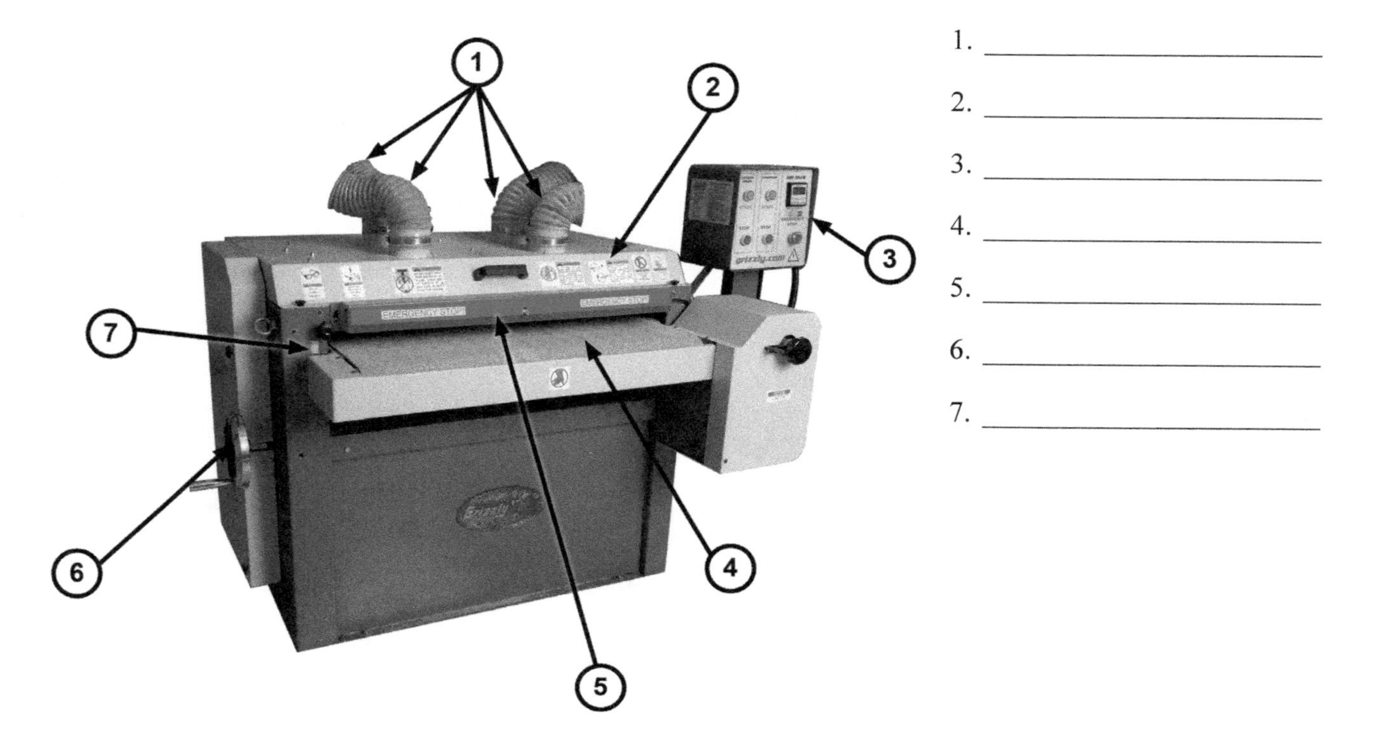

1. ______________________
2. ______________________
3. ______________________
4. ______________________
5. ______________________
6. ______________________
7. ______________________

SAFE OPERATIONAL PROCEDURES

1. Keep bystanders and yourself away from the infeed and outfeed ends when a material is fed into the sander.
2. Keep your hands away from the sanding drums during operation.
3. Keep fingers away from the conveyor and the underside of the material during sander and conveyor operation.
4. Adjust the conveyor feed rate and sanding drum height, so your first pass has light pressure, to avoid overloading the sander.
5. Never force the workpiece into the sander to avoid kickbacks.
6. Shut the sander down, let the drums come to a complete stop, and disconnect power before you service, adjust, troubleshoot, or leave the machine unattended.
7. Never attempt to clear a jammed workpiece while the sander is running.
8. Replace the sandpaper when it is worn, and only use undamaged sandpaper.

9. Inspect the workpiece for nails, staples, knots, imbedded stones, and other material that could be dislodged and thrown from the machine during sanding operations.
10. Do not sand if there is any doubt about the stability or integrity of the material.
11. Placing fingers between material and conveyor can result in pinching injuries, or possibly getting trapped and pulled into sanding area of machine. DO NOT place fingers under bottom of workpiece while feeding it into sander.
12. Rotating sandpaper can remove a large amount of flesh quickly. Keep hands away from rotating sanding drum(s) during operation. Never touch moving sandpaper.
13. This sander is designed to sand only natural wood materials or man-made products made from natural wood fiber. DO NOT sand any metal materials.
14. Never sand more than one material at a time.
15. Never sand material smaller than 1/8" thick x 9" long.
16. Do not sand thin material by using a "sled" (another board) under the material.
17. Never adjust the conveyor belt tracking when the sanding drums are engaged.
18. Always feed material against the rotation of drum

GENERAL SAFETY PRACTICES

1. Wear approved eye protection, hearing protection, and proper clothing. Tie up loose hair and remove loose jewelry.
2. Do not operate the machine without the instructor's permission, or without instructor supervision.
3. Keep all clothing, and long hair away from all sander moving parts.
4. Remove tools from machine, tools left on machinery can become dangerous projectiles upon startup.
5. Keep proper footing and balance at all times when operating machine. Do not overreach! Avoid awkward hand positions that make material control difficult or increase the risk of accidental injury.
6. Keep guards and covers in place to reduce accidental contact with moving parts or flying debris. Make sure they are properly installed, undamaged, and working correctly BEFORE operating machine.
7. Keep this machine in correct adjustment and properly serviced.
8. Do not force the machine. It will do the job safer and better at the rate for which it was designed.
9. Regularly inspect the machine for damaged, loose, or mis-adjusted parts—or any condition that could affect safe operation. DO NOT operate the machine with damaged parts!

COMPLETION QUESTIONS

1. Do not sand if there is any doubt about the _______________ or ________________ of the workpiece.
2. Do not _____________ machine. It will do the job safer and better at the rate for which it was designed.
3. Always feed material _________________ the rotation of drum.
4. Keep _______________ away from the conveyor and the underside of the workpiece during sander and conveyor operation.
5. Keep proper ______________and _______________ at all times when operating machine.
6. Rotating sandpaper can remove a large amount of _______________ quickly.
7. Keep _______________ and long hair away from all moving parts.
8. Do not sand _______________ material by using a "sled" (another board) under the workpiece.
9. Never sand more than _______________ workpiece at a time
10. Never sand material smaller than _______________ thick x _______________ long.

EDGE SANDER

PART IDENTIFICATION

Identify the circled parts on the edge sander illustrated below.

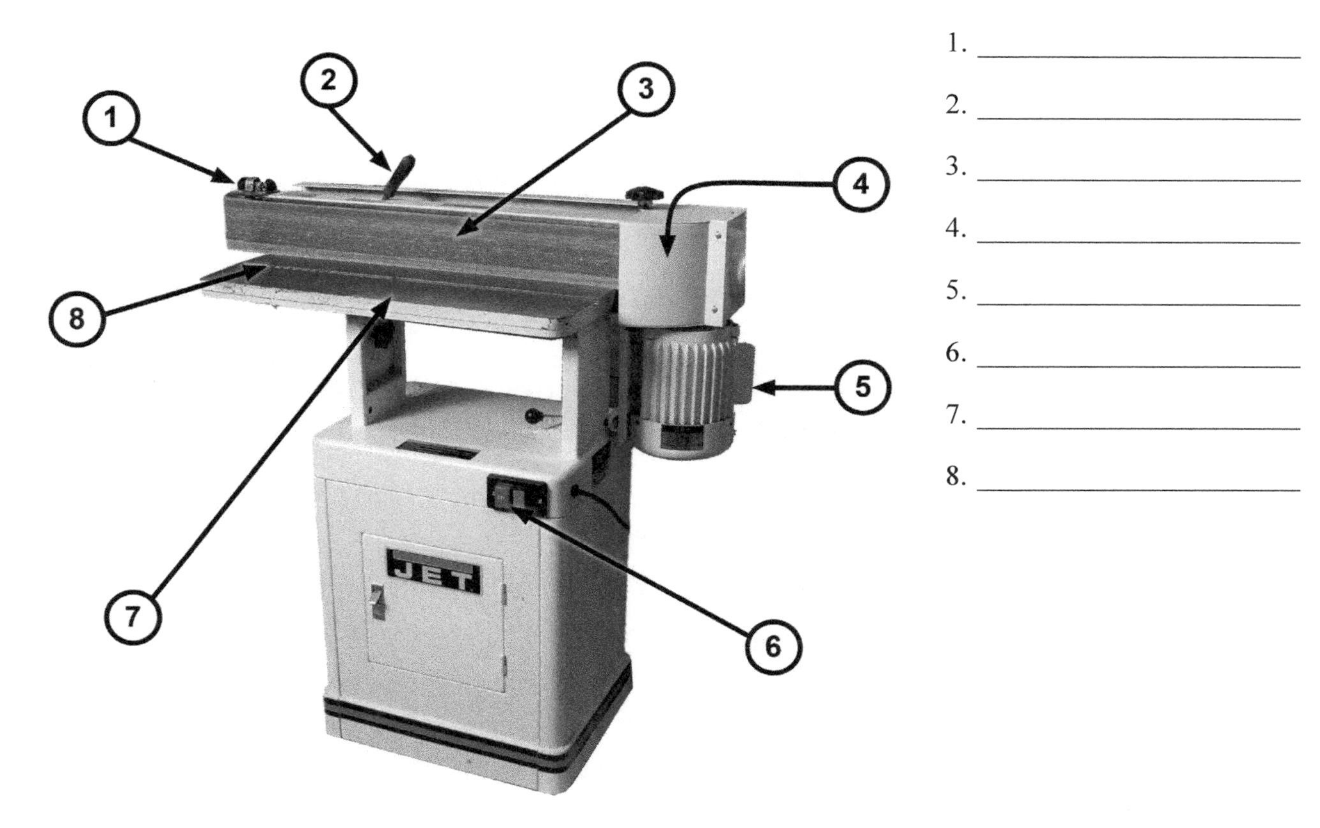

1. ______________
2. ______________
3. ______________
4. ______________
5. ______________
6. ______________
7. ______________
8. ______________

SAFE OPERATIONAL PROCEDURES

1. Using the belt sander
   a. Identify how much material needs to be removed by marking the material and sanding up to that line to avoid over sanding.
   b. Hold onto the material when starting up the machine, ensure the material will not come into contact with the belt until you are prepared and the machine is up to full speed.
   c. Lay the material flat on the table, using firm pressure, move the material into the belt and move slowly against the rotation of the belt.
   d. When trying to maintain a square edge a miter gauge can be used to hold the material 0-90 degrees from the belt while moving the material against the rotation of the belt.
   e. When finished with the sanding process using firm pressure slide the material toward the front edge of the table away from the belt, then shut down the power and stay with the machine until the belt has coasted to a stop.

f. When edge sanding the cover/guard should be closed over the spindle sander end to minimize potential risks.

2. Using the spindle sander
   a. Identify how much material needs to be removed by marking the material and sanding up to that line to avoid over sanding.
   b. Hold onto the material when starting up the machine, ensure the material will not come into contact with the belt until you are prepared and the machine is up to full speed.
   c. Lay the material flat on the table, using firm pressure, move the material into the belt and move slowly against the rotation of the spindle.
   d. With concave shapes it is important to hold the material firmly as you sand around the curved profile, a rocking motion will allow you greater control and pressure of the material instead of trying to move the material linearly.
   e. When finished with the sanding process using firm pressure slide the material toward the front edge of the table away from the spindle, then shut down the power and stay with the machine until the belt has coasted to a stop.

GENERAL SAFETY PRACTICES

1. Wear approved eye protection, hearing protection, dust protection such as a mask, and proper clothing. Tie up loose hair and remove loose jewelry.
2. Do not operate the machine without the instructor's permission, or without instructor supervision.
3. Always move the material along the belt instead of holding in place to reduce the burning of the material and wear on the belt.
4. Always move the material against or in opposition to the rotation of the belt to avoid having the material pulled out of your hands.
5. Only use the sander in a well-ventilated space.
6. Keep fingers away from the belt, paying special attention to keep knuckles clear of the belt at all times.
7. As the belt wears down the table should be raised or lowered to allow the belt to wear evenly.
8. An abrasive cleaning stick or belt cleaning stick should be used frequently to clear the built-up debris embedded into the belt. Use the cleaning stick when the machine is running just as you would sand material, the rubber will help remove the debris and extend the life of the belt.

COMPLETION QUESTIONS

1. A ________________ ________________ should be used to maintain square edges from 0-90 degrees.
2. You should ________________ the material before beginning sanding to avoid over sanding the material.

3. Material should be placed flat on the ________________ for the entire sanding process.
4. You should move the material constantly, against the rotation of the belt to prevent ________________.
5. A belt cleaning stick will ________________ the life of the belt and minimize the burning of material.
6. The sander should only be used in a well-ventilated space while you are wearing a ________________ that protects you from breathing in fine dust particles.
7. Your instructor will adjust the table up and down to ensure the ________________ wears evenly, to extend the life of it.
8. When operating the edge sander, PPE such as closed-toe shoes, ________________ ________________, and a mask is essential.
9. The cover/guard should always cover the ________________ portion of the edge sander when not in use.
10. A ________________ motion will allow you greater control and pressure when spindle sanding.

GAS FORGE

PART IDENTIFICATION

Identify the numbered parts of the gas forge illustrated below.

1. ______________________
2. ______________________
3. ______________________
4. ______________________
5. ______________________
6. ______________________
7. ______________________
8. ______________________
9. ______________________
10. ______________________
11. ______________________

SAFE OPERATIONAL PROCEDURES

1. Ignition instructions for spark ignited forge (follow manufacturer's instructions):
   a. If the forge has been in operation, always wait at least five (5) minutes between shutdown and starting up of the forge.
   b. Set air control halfway between open and closed positions. Set gas control to closed position.
   c. Using the lid handle, swing the lid toward the back side of the furnace so that it is not over the top slot.
   d. Depress and release the start button. The blower motor will start running, the red light will come on, and if the room is not too noisy, the spark plug igniter "buzzing" will be heard. The igniter will stay on for about 1-1/2 minutes, so the starting cycle must be completed during this time.

e. Depress and **hold in** the igniter button. Slowly, turn the gas control toward the open position until the burners ignite. Then turn the control slightly past this position to obtain a steady roar from the burners. After about twenty (20) seconds, the red light will go out and the ignition button can be released. If the lighting cycle was not completed in 1-1/2 minutes, a thermal relay will shut off the gas and the spark igniter. Push the "stop" button, wait five (5) minutes, then repeat steps b, c, d, and e to start the forge.

f. After the forge has been started, adjust the gas control to give a sharp tail of flame that extends just above the top of the forge. Work can be placed in the flame. The work rack at the front of the furnace can be slid out to support the work. The lid should be centered over the top slot, using the lid handle.

g. To increase the amount of gas, turn the gas control toward the open position to get a higher flame. Then turn the air control handle toward the open position to obtain the sharp tail of flame. Repeat these steps until the desired or maximum gas input is reached.

h. To decrease the amount of gas, turn the gas control handle toward the closed position until the sharp tail of flame almost disappears. Then turn the air control toward the closed position until the tail of flame reappears. Repeat this procedure until the desired or minimum gas input is reached.

i. If the gas is turned too high causing a high, lazy flame or too low causing an intermittent flame, the thermocouple may cool down and shut off the burners. Push the stop button, wait five (5) minutes, and restart the forge.

2. Shutting down the spark ignited forge:

   a. To shut down the forge, turn gas control to the closed position and push the stop button.

   b. Do not attempt to speed the cooling process in the fire box by restarting the blower. The firebrick should cool slowly and evenly to prevent cracking.

3. Ignition instructions for the manually ignited forge:

   a. Pivot lid back until it no longer covers top slot.

   b. Turn solenoid gas control switch to "off" position.

   c. Plug in blower motor; blower motor should start running.

   d. Turn valve in gas line to 1/2 open position.

   e. Ignite a fuel-soaked wick held in tongs; hold burning wick in fire box.

   f. Keeping hands, face, and body away from top slot, turn solenoid gas control switch to "on" position. When forge has ignited, place remaining wick in a metal container and cover with a metal lid to extinguish flame.

   g. Adjust flame height to top of the forge with the gas line valve. Minor air adjustments may be made by rotating the disc which is located in front of the blower air inlet.

4. Shutting down the manually ignited forge:

   a. Turn solenoid gas control switch to "off."

   b. Close the gas line valve.

   c. Unplug the cord to the blower motor. Failure to stop blower may result in unequal cooling of firebrick and cracking.

GENERAL SAFETY PRACTICES

1. Wear approved eye protection, hearing protection, and proper clothing. Tie up loose hair and remove loose jewelry.
2. Do not operate the machine without the instructor's permission, or without instructor supervision.
3. Keep area near forge cleared of everything except tong rack and bucket of cool water to prevent accidental tripping of individuals using the forge.
4. All flammable objects and materials must be kept at a safe distance from the hot metal working area.
5. Adjust lid to proper height prior to igniting forge.
6. Wear safety shield over face to protect face and eyes from flame, sparks, and heat.
7. Wear heavy leather gloves when working with hot metals. Exercise great care in handling metal even with gloves.
8. Never wear oily or loose-fitting clothing near forge.
9. Use only the right tongs for the job.
10. Turn on ventilation fan before igniting forge and do not turn off until forge has been turned off.
11. Have fire extinguisher, fire blanket, and first aid kit readily available.

COMPLETION QUESTIONS

1. Before starting the gas forge ignition procedure, the ____________________ ____________________ should be turned on.
2. At the start of the ignition procedure, the gas control should be ____________________ and the air control ____________________ ____________________.
3. In shutting down the forge, the ____________________ control should be closed first.
4. The height of the flame should be just above the ____________________.
5. Do not reignite the forge within ____________________ minutes of shutdown.
6. The proper flame shape resembles a ____________________ ____________________.
7. Flame height is adjusted with the ____________________ ____________________ ____________________.
8. Minor ____________________ adjustments may be made by rotating the disc located near the blower inlet.

9. In shutting down the manually ignited forge, first turn the ________________________ ________________________ ________________________ to the “off” position, close the ________________________. ________________________ ________________________, and then stop the ________________________.

10. Failure to stop the ________________________ could result in cracking of the firebrick.

HORIZONTAL METAL-CUTTING BAND SAW

PART IDENTIFICATION

Identify the numbered parts of the horizontal metal-cutting band saw illustrated below.

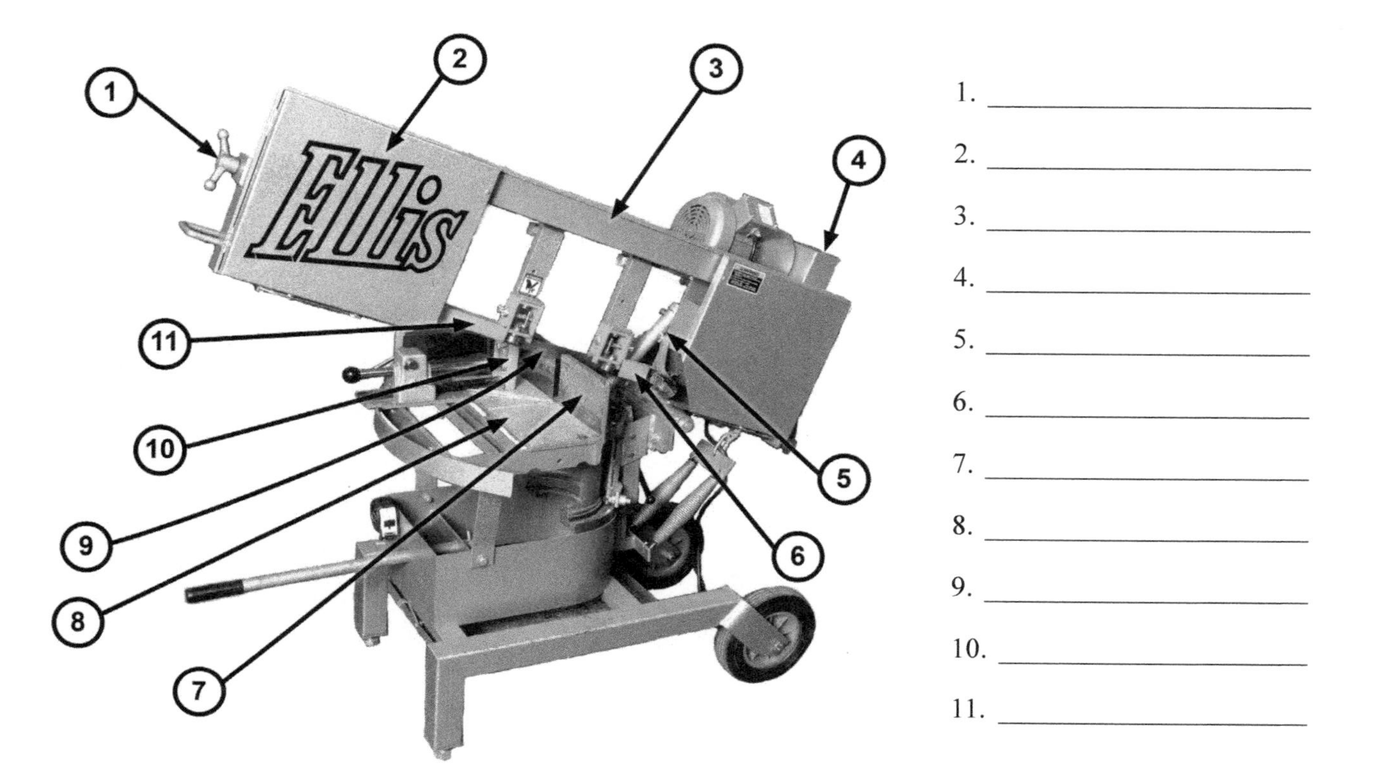

SAFE OPERATIONAL PROCEDURES

1. Replacing blade:
   a. Select a blade that is sharp and in good condition with 14-18 teeth per inch for thin wall tubing.
   b. Place the blade in the saw with the exposed teeth pointing toward the motor.
   c. Tighten the blade until it is snug; do not over-tighten.
   d. When a worn or broken blade is replaced, the material being cut should be turned over. A new blade is thicker and will be damaged if allowed to enter an old saw kerf.
2. Placing material in the vise of a horizontal metal-cutting band saw:
   a. Clamp all material firmly on the vise to prevent breaking the blade.
   b. Clamp angle iron in the vise with the legs down.
   c. Clamp rectangular tube and bar stock material with the widest side toward the blade.
   d. Support long material with roller stands, not a partner.

   e. When cutting short material, a block of equal width can be placed in the opposite end of the vise jaws. This will allow the vise to grip the material tighter.
   f. Never use a piece of pipe or a wrench as a lever to help tighten the vise.
   g. Use a cut-off gauge when cutting short pieces of material the same length.

3. Sawing:
   a. Properly clamp the material in the vise.
   b. Lower the blade and check for accuracy. The blade should not contact the material unless the saw is running. The cut should be made on the waste material side.
   c. Move the adjustable blade guide as close to the work as possible without touching.
   d. Raise the blade, start the saw, and then lower the blade gently onto the work.
   e. Do not force the saw into the work.
   f. Do not attempt to break off the material before the cut is complete.
   g. Wait for the machine to automatically shut off at the end of the cut.
   h. Release the metal from the vise.
   i. Clean the area before leaving. Remove all scrap from the floor.

4. Lubricant
   a. Some saws also have a pump that directs cutting fluid onto the blade to help keep it and the work piece cool which aids in a faster and cleaner cut as well as prolonging the life of the blade. If your saw has this feature:
      (1) Never operate the saw without the fluid running.
      (2) Adjust the lubricant flow so that it doesn't splash out of the machine.
      (3) Aim the lubricant on the surface of the blade just in front of the work piece.
      (4) Ensure there are no obstructions in the cutting area stopping fluid from flowing back to the reservoir.
      (5) Check the screen/filter on the pump for obstructions and clean often.

GENERAL SAFETY PRACTICES

1. Wear approved eye protection, hearing protection, and proper clothing. Tie up loose hair and remove loose jewelry.
2. Do not operate the machine without the instructor's permission, or without instructor supervision.
3. Keep hands a reasonable distance from the blade when cutting.
4. Keep the adjustable blade guide as close as possible to the material.
5. Be sure the metal to be sawed is held securely in the vise.
6. Make sure the pulley shield is always in place before use.
7. Do not wear loose-fitting clothes or a wrist watch.

8. If the blade breaks, do not attempt to stop the machine; it will stop automatically when the feed control finishes lowering the saw blade.
9. Shut off main power switch and unplug before making adjustments or repairs on the machine.
10. Allow the blade to feed in the work slowly when starting a cut.
11. Do not force the speed of cutting by applying pressure.
12. Support long material with roller stands and not a partner.
13. Do not leave saw running unattended.
14. Remove the burr left by the saw by filing or grinding.
15. After the saw has stopped, use a brush, not bare hands, to remove chips from the machine.
16. Never use compressed air to clean saw area.
17. Remove scrap material from the floor.

COMPLETION QUESTIONS

1. When replacing the blade in a horizontal metal-cutting band saw, the teeth should be pointing toward the ___________________________.
2. When a worn or broken blade is replaced, the material being cut should be ___________________________ ___________________________.
3. All material should be firmly clamped in the ___________________________ to prevent breaking the blade.
4. Angle iron should be clamped in the vise with the legs ___________________________.
5. The widest side of rectangular material should be toward the ___________________________.
6. A ____________________ ______________________ gauge can be used when cutting off several pieces of material the same length.
7. The cut should be made in the ______________________ material.
8. Do not ___________________________ the saw into the work.
9. Do not break off the ___________________________ before the cut is complete.
10. Before starting the saw, the adjustable blade ___________________________ should be moved as closely as possible to the work.

IRONWORKER/METALWORKER

PART IDENTIFICATION

Identify the circled parts on the ironworker/metalworker illustrated below.

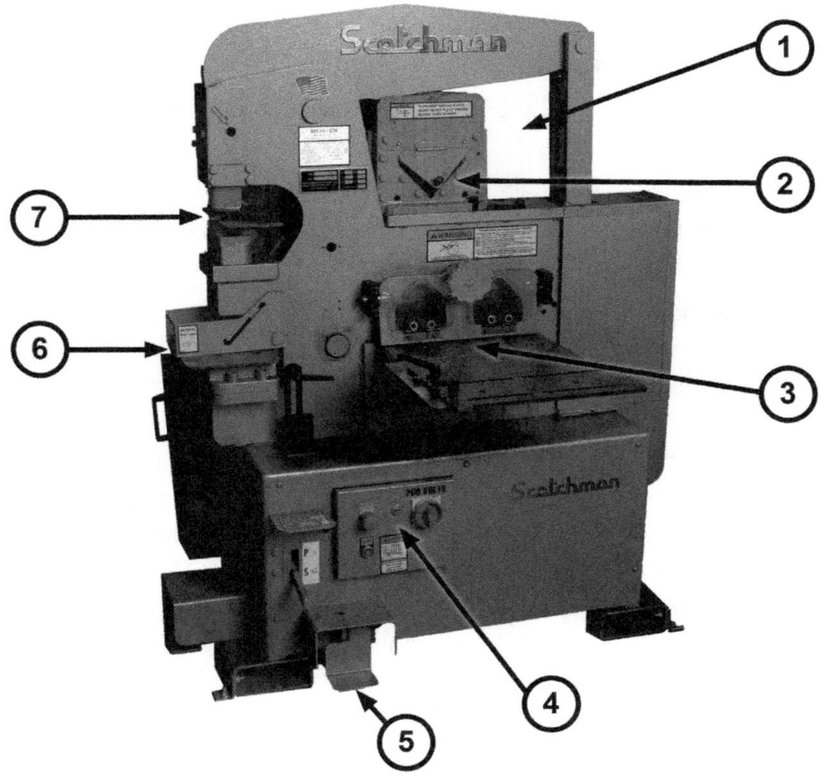

1. ______________________________
2. ______________________________
3. ______________________________
4. ______________________________
5. ______________________________
6. ______________________________
7. ______________________________

SAFE OPERATIONAL PROCEDURES

1. Shearing:
   a. Prior to operation:
      (1) Ensure the punching and notching areas are clear.
      (2) Ensure the area below and around the foot pedal are clear.
   b. Operation:
      (1) Turn on the power.
      (2) Place the material between the blades of the shear.
      (3) Adjust the guard to the minimum setting that still allows a safe clearance, never operate the machine without the guard in place.
      (4) Move away from the shear blades, do not touch the machine or the material, use clamps to hold pieces, when necessary, depress the foot pedal to begin the shearing operation.

(5) When the shearing process is complete lift your foot from the foot pedal, wait for the shearing blade to automatically return to the top stroke position for the next operation.

(6) Cut off material will fall down the material slide on the backside of the machine, stay clear until all movement has ceased.

(7) Clear away cut off material debris as needed to keep the work area is clear.

2. Punching:

   a. Prior to operation:

      (1) Ensure the shearing and notching areas are clear.

      (2) Check the die, the nut holding the punch and the stripper frequently to ensure they are tight.

      (3) Do not punch anything thicker than the diameter of the punch.

      (4) Always punch full holes, not partial holes, the side thrust from cutting a partial hole could damage the cutter.

      (5) Clear the area under the beam and punch ram.

   b. Operation:

      (1) Turn on the power.

      (2) Check that the punch and die are aligned before beginning the operation, and throughout the day.

      (3) Place the material in position to be punched, do not stack material, punch only 1 piece at a time.

      (4) Move away from the punch, do not touch the machine or the material, the use of clamps is not necessary when punching material, as the punch applying pressure will act as a clamp during the operation.

      (5) Depress the foot pedal to begin the punching operation.

      (6) When the process is complete lift your foot from the foot pedal, wait for the punch to automatically return to the top stroke position for the next operation.

      (7) Clear away cut off material debris as needed to keep the work area clear.

3. Notching:

   a. Prior to operation:

      (1) Ensure the punching and notching areas are clear.

      (2) Ensure the area below and around the foot pedal are clear.

   b. Operation:

      (1) Turn on the power.

      (2) Place the material between the blades of the notcher.

      (3) Adjust the guard to the minimum setting that still allows a safe clearance, never operate the machine without the guard in place.

      (4) Move away from the notcher, do not touch the machine or the material, use clamps to hold pieces, when necessary, depress the foot pedal to begin the notching operation.

      (5) When the notching process is complete lift your foot from the foot pedal, wait for the

notcher blade to automatically return to the top stroke position for the next operation.

(6) Clear away cut off material debris as needed to keep the work area is clear.

4. Bending Break:
   a. Prior to operation:
      (1) Ensure the punching and notching areas are clear.
      (2) Ensure the area below and around the foot pedal are clear.
   b. Operation:
      (1) Turn on the power.
      (2) Place the material between the dies for breaking material.
      (3) Always load material from the front table side of the machine.
      (4) Use clamps to hold small pieces, when necessary, depress the foot pedal to begin the breaking operation.
      (5) When the breaking process is complete lift your foot from the foot pedal, wait for the bending dies to automatically return to the top stroke position for the next operation.

GENERAL SAFETY PRACTICES

1. Wear approved eye protection, hearing protection, and proper clothing. Tie up loose hair and remove loose jewelry.
2. Do not operate the machine without the instructor's permission, or without instructor supervision.
3. Use safety guards, material hold-downs and punch stripper supplied with your ironworker/ metalworker. Removal, modification or improper use of these safety devices may result in serious injury.
4. Always keep hands and body away from pinch points and any moving parts during operation.
5. Stay within rated capacity when using the ironworker/metalworker.
6. The ironworker/metalworker is only recommended for only mild steel. Harder materials can damage the machine.
7. Do not stack material; cut, punch, notch, or bend only one piece at a time.
8. Always be ready to lift your foot off the foot pedal to stop a mistake from seriously injuring you.
9. Maintain a clean machine. Remove any obstructions, slugs, cut-offs and fillings from the work area.
10. Adequately support and clamp the steel material being worked.
11. Keep the area under and around the foot pedal clear.
12. Remove all scrap and tools from the iron worker to avoid injury or damage.
13. Use proper shut down procedures before changing punches, blades or shims.

14. Turn your ironworker/metalworker off when not in use—never leave a powered ironworker/ metalworker unattended.
15. Periodically check tooling for wear. Replace worn tooling according to manufacturer's guidance.
16. Periodically clean your ironworker/metalworker with a compressed air nozzle and soft cloth. Remove filings, dirt, dust and grime from working surfaces.
17. Periodically inspect your ironworker/metalworker to ensure that all switches, wires and plugs are in good operating condition.

COMPLETION QUESTIONS

1. Ensure that all switches, wires and plugs are in ______________ operating condition.
2. Rings and any loose ______________ must not be worn while operating any machine.
3. Check the punch and die for ______________ prior to punching the first hole.
4. Make sure all guards are in ______________.
5. Never put your hands near a hazardous area or ______________ ______________ of the machine.
6. Use proper shut down ______________ before changing punches, blades or shims.
7. Keep the area under and around the foot ______________ clear.
8. Stay within rated ______________ when using the ironworker/metalworker.
9. Do not attempt to punch material thicker than the ______________ of the punch.
10. At the conclusion of the stroke, remove your ______________ from the pedal of the machine.

JOINTER

PART IDENTIFICATION

Identify the numbered parts of the jointer illustrated below.

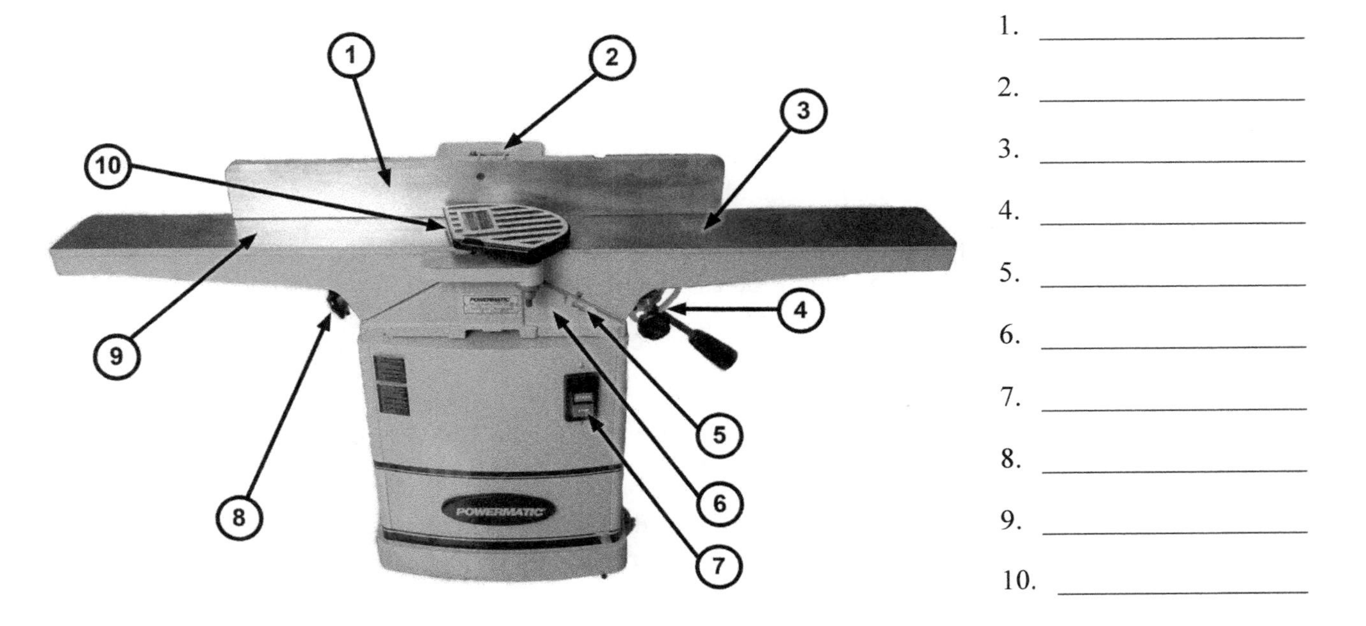

1. ____________________
2. ____________________
3. ____________________
4. ____________________
5. ____________________
6. ____________________
7. ____________________
8. ____________________
9. ____________________
10. ____________________

SAFE OPERATIONAL PROCEDURES

1. Keep the guard covering the cutterhead in place and in working order at all times.
2. The jointer's main function is to square or straighten the edge of material and it should not be used as a surfacer or planer.
3. Keep the knives sharp and properly adjusted. Dull knives are more likely to cause a kickback. Knives out of adjustment (either cutting on an angle or one knife adjusted lower or higher than the other) will cause improper cutting and possibly a safety hazard.
4. When the knives are replaced, all knives should be replaced and be sharpened alike as a set. Due to the speed of the cutterhead (approximately 6,000 RPM), knives must be in balance. Most cutterheads will have three knives.
5. The depth of cut is adjusted with the front infeed table adjusting hand wheel or lever. The machine should never be set to remove more than 1/8" of material in a single pass for softwoods or 1/16" of material in a single pass for hardwoods.
6. The depth of cut scale should be checked and readjusted to zero each time new blades are added or after adjustments to the cutterhead.

7. To work properly, the rear outfeed table must be exactly level with the knives in the cutterhead. Check by laying a straightedge over the table and cutterhead and then adjusting the table up or down as needed. Only a qualified person should make this adjustment.

8. The fence should be moved back and forth to different positions periodically so as not to cut at the same position on the cutterhead. Check the angle of the fence with a square, and adjust fence pointer or scale to the zero point.

9. Bevel cuts may be made by adjusting the fence to the proper angle. Bevel cuts at sharp angles are more dangerous as the material will tend to slide down the fence and float over the cutterhead.

10. Check the direction of the grain of the material to be certain the knives are cutting with the grain and not against it. A beginner should not attempt to joint end grain.

11. Always have a push-stick accessible so it can be used on short pieces or at the end of a board so as to keep hands as far from cutterhead as possible.

12. If jointing long pieces, use a roller support or have someone support the piece after it passes over the rear outfeed table.

13. Do not stand directly in line with the tables in case of a kickback. Also see that others are not in line with the jointer when in use.

14. The cutterhead turns down toward the front infeed table; therefore, the revolving cutterhead will have a tendency to throw the work back toward the infeed table.

15. Do not attempt to joint material less than 1/2” thick, 3” wide, or pieces shorter than 12” in length, without using a push stick.

16. Feed material into the jointer slowly. Make certain the material is pressed firmly against the table and fence surfaces as it is moved across the work surface.

17. Turn off the power and allow the cutterhead to stop rotating before leaving the area.

18. Test the squareness of edges with a square and by stacking pieces of jointed material on their edges.

GENERAL SAFETY PRACTICES

1. Wear approved eye protection, hearing protection, and proper clothing. Tie up loose hair and remove loose jewelry.

2. Do not operate the machine without the instructor’s permission, or without instructor supervision.

3. Use only sharp knives.

4. Keep floor and area around jointer clean and free of scraps and debris.

5. Do not make any adjustments with machine running.

6. Double check all adjustments before turning on power.

7. Do not attempt to joint or edge material containing nails, paint, severe knots, or other defects.
8. Always keep the guard in place and working freely.
9. Do not allow hands to pass directly over the cutterhead.
10. Do not talk to others while operating the jointer.

COMPLETION QUESTIONS

1. Material shorter than ______________________ inches should not be run on the jointer.
2. Do not allow hands to pass directly over the ______________________.
3. The depth of cut is adjusted by moving the ______________________ ______________________ table up or down.
4. Bevel cuts may be made by adjusting the ______________________ to the proper angle.
5. Material should be at least ______________________ thick to be run on the jointer.
6. A ______________________ ______________________ should be used for short material and in finishing the end of a long material as it passes over the cutterhead.
7. The ______________________ ______________________ table should only be adjusted by a qualified person.
8. All ______________________ should be changed as a set on the cutterhead when dull to keep it in ______________________.
9. The material should be fed into the jointer so the machine is cutting ______________________ the grain and not ______________________ the grain.
10. The jointer should never be operated with the ______________________ removed.

LASER ENGRAVER

PART IDENTIFICATION

Identify the circled parts on the laser engraver illustrated below.

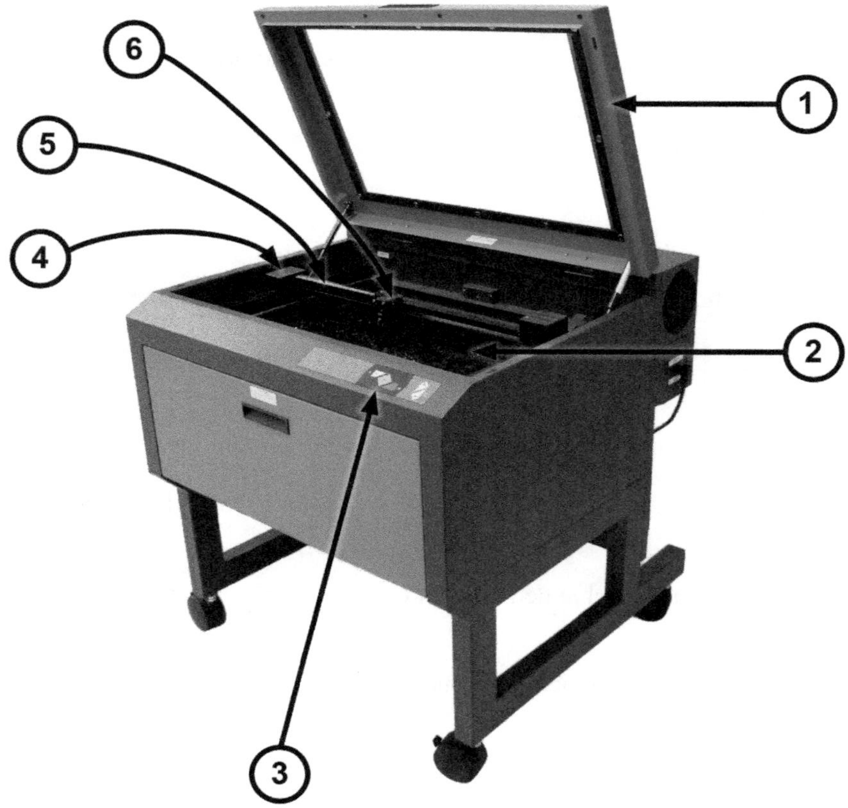

1. ____________________
2. ____________________
3. ____________________
4. ____________________
5. ____________________
6. ____________________

SAFE OPERATIONAL PROCEDURES

1. Never operate the laser with the lid or doors open, overriding the lock system exposes the user to potentially harmful fumes and radiation from the laser.
2. Ensure the laser head will clear any material hold down devices being used.
3. Verify that the workpiece is made of an approved material and is safe to use and has appropriate geometry for use on the particular laser cutter.
4. Identify the appropriate laser settings for the material and intended operations.
5. Remove any protective coatings from the workpiece (these tend to cause flare-ups).
6. Confirmed location and condition of the nearest exit and fire extinguisher.
7. Turn on laser.

8. Ensure that laser starts appropriately and that the cutting head and laser bed are in correct home positions and have stopped all motion.
9. If you are cutting through material ensure the cutting bed is installed.
10. Open cover and ensure that there is no debris or objects on laser bed.
11. Place workpiece on laser bed in correct orientation/location.
12. Ensure/start laser exhaust system and confirm proper exhaust flow.
13. Close protective cover and confirm all interlocks have been satisfied (no flashing red lights).
14. Start laser operations.
15. Observe all laser operations but don't stare at cutting beam.
16. Ensure that there are no continued flare-ups during cutting and that all flames self-extinguish when the laser is not engaged.
17. At completion of laser operation, allow workpiece to cool and exhaust fan to purge fumes for several minutes according to local shop practices.

GENERAL SAFETY PRACTICES

1. Wear approved eye protection, hearing protection, and proper clothing. Tie up loose hair and remove loose jewelry.
2. Do not operate the machine without the instructor's permission, or without instructor supervision.
3. Always refer to the operator's manual to determine if a material is safe to laser engrave/cut.
4. Always refer to the operator's manual to determine recommended settings for the material being laser engraved/cut.
5. Never leave the laser unattended while in operation.
6. Always have a cold damp rag readily available. In case of fire open the lid causing the machine to stop immediately and place a cold damp cloth on the material.
7. Always use the exhaust/filtration system with the laser engraver this may be a built in to the wall or a standalone unit hooked up to the machine. Do not use the same exhaust system that woodworking equipment/machines run on due to the potential fire hazards.
8. Do not look directly at the laser beam.
9. Do not operate the laser unless all covers are in place and in good repair.
10. Laser cutting and engraving will generate vapors and fumes from the substrate.
11. Never place hands or other body parts inside laser enclosure during any operations.
12. Fresh cuts on work piece may produce burrs and other sharp edges.

13. Students may attempt to "break out" partially cut thru material with the high potential for small sharp object to fly into unprotected eyes or cause cuts on fingers and hands.
14. Inspect the tool for damage prior to use.
15. Verify all guards are in place and adjusted properly.
16. Do not bypass any safety devices.
17. Report any malfunction or damage to the instructor.
18. Understand the uses, limitation, and hazards of the laser cutter/engraver.

COMPLETION QUESTIONS

1. Never operate the laser with the ________________, overriding the lock system exposes the user to potentially harmful fumes and radiation from the laser.
2. Verify that the workpiece is made of an ________________ and is safe to use and has appropriate geometry for use on the particular laser cutter.
3. Remove any protective ________________ from the workpiece (these tend to cause flare-ups).
4. Identify the appropriate ________________ for the material and intended operations.
5. Observe all laser operations but don't stare at ________________.
6. Never leave the laser ________________ while in operation.
7. Always use the ________________ with the laser engraver; this may be a built in to the wall or a stand alone unit hooked up to the machine.
8. Do not operate the laser unless all ________________ are in place and in good repair.
9. Open cover and ensure that there is no ________________ or objects on laser bed.
10. Wear approved ________________, hearing protection, and proper clothing. Tie up loose hair and remove loose jewelry.

MEDIA BLASTER

PART IDENTIFICATION

Identify the circled parts on the media blaster illustrated below.

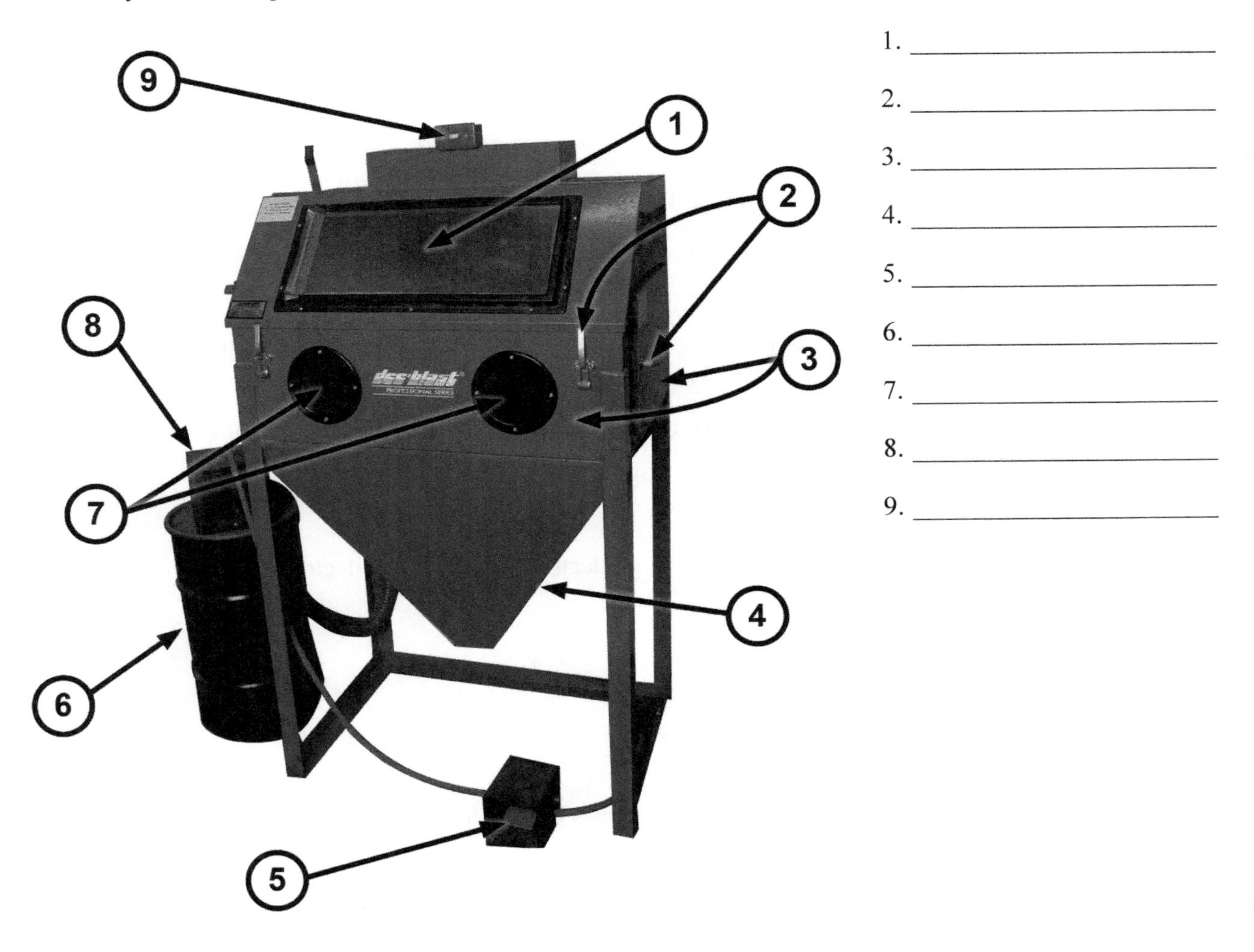

1. ____________________
2. ____________________
3. ____________________
4. ____________________
5. ____________________
6. ____________________
7. ____________________
8. ____________________
9. ____________________

SAFE OPERATIONAL PROCEDURES

1. Check air lines and hoses for weak or worn condition before each use.
2. Make sure all the air line connections are tight and secure before use.
3. Check the media level before each use to ensure there is enough media to complete the job. However, do not add too much media, this may cause the cabinet to not seal correctly. Always use approved blaster media.
4. Always check the cabinet dust seals, look for any damage, aging, deformation and air leaks resulting in air loss.
5. Ensure proper lighting around and inside the media blaster.

6. Ensure cabinet door is closed and all the cabinet components are sealed properly before using.
7. Do not exceed the maximum operating pressure of the blasting equipment.
8. Maintain a safe perimeter around the blasting area to ensure other's safety.
9. Do not use the blaster around combustibles or flammables that can explode or burn quickly. Some abrasives create sparks which will cause a fire.
10. During operation, do not expose your hands or skin directly in the line of the blast nozzle.
11. Do not point the blaster nozzle at anyone or objects.
12. Make sure to wear gloves when handling freshly blasted parts.
13. Disconnect the cabinet from the air supply before changing accessories or attempting to install, service, relocate or perform any maintenance.
14. When you finish, clean all the abrasives out in order to prevent a breakdown for the next operator.
15. Washing hands is recommended after using products being blasted, they can include harmful chemicals.
16. Discharge water buildup in the air hose daily.
17. Clean dust collector filters as needed, failure to clean will cause the collector motor to overheat.

GENERAL SAFETY PRACTICES

1. Wear approved eye protection, hearing protection, and proper clothing. Tie up loose hair and remove loose jewelry.
2. Do not operate the machine without the instructor's permission, or without instructor supervision.
3. Do not use blast media containing free silica. Silica can cause silicosis (cancer) or other related respiratory damage.
4. Do not operate this equipment without wearing OSHA approved respiratory protection.
5. All blast equipment operators and personnel entering the vicinity of the blast operation must use respiratory protective equipment that meets OSHA regulations.
6. Keep all media dry to avoid clogs.
7. Do not operate this equipment in a manner other than its intended application, and do not modify or alter any abrasive blaster, blast equipment, or controls.
8. Do not use this equipment with damaged components. Periodically check all valves, hoses, and fittings to see that they are in good condition.

9. To minimize the risk of electrocution, fire, or equipment damage, installation work and electrical wiring must be done by an electrician or qualified service person.
10. To avoid personal injury and property damage, study the service manual before assembling, operating, or servicing the sand blasting cabinet; and read all SDS of the materials and chemicals you will be in contact with.
11. Clean the media blaster machine or parts with a soft cloth, do not use solvents to clean.

COMPLETION QUESTIONS

1. During operation, do not expose your ________________ or ________________ directly in line of the blaster nozzle.
2. Ensure that the cabinet is ________________ properly before starting the blaster.
3. Do not point the blaster ________________ at anyone.
4. Do not exceed the maximum operating ________________ of the blasting equipment.
5. Disconnect the cabinet from the ________________ ________________ before attempting any service or maintenance.
6. Check hoses and lines for damage, make sure all ________________ are secure before use.
7. Do not add too much ________________, this may cause the sealed cabinet door to leak.
8. Failure to clean the dust collector filters may cause the ________________ ________________ to overheat.
9. Do not operate this equipment without wearing ________________ approved respiratory protection.
10. The media blaster abrasive dust may contain chemicals known to cause ________________.

METAL BELT GRINDER

PART IDENTIFICATION

Identify the numbered parts of the metal belt grinder illustrated below.

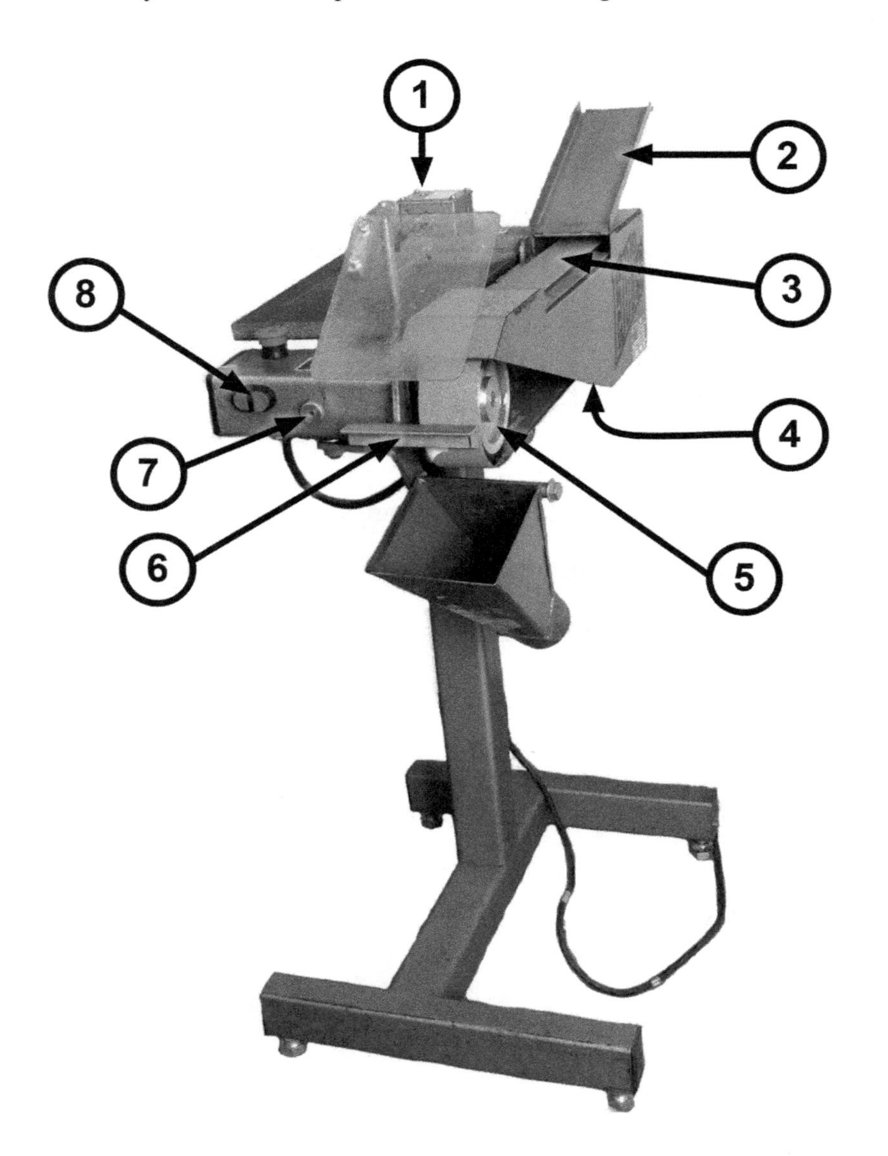

1. ______________________
2. ______________________
3. ______________________
4. ______________________
5. ______________________
6. ______________________
7. ______________________
8. ______________________

SAFE OPERATIONAL PROCEDURES

1. Check the belt to see if it is torn or otherwise damaged.
2. Make sure the belt runs on the center of the pulleys.
3. Select a new belt if needed and mount it on the pulleys.

4. Check the belt direction before attempting to grind. The belt is spliced for operation in only one direction. If the belt is mounted wrong, it will tear.
5. Adjust the table so it clears the belt by 1/8".
6. Check the belt for tightness. If the belt is too loose, it will pile up in front of the work. If the belt is too tight, there will be unnecessary wear on the bearings.
7. If the belt does not run centered on the pulleys, the idle pulley will need to be tilted to make one side of the belt tighter than the other. The belt will run toward the looser side of the pulleys.
8. Start the motor before applying the work to the belt and let it come up to full speed.
9. Feed metal slowly; do not force it. Apply firm, but not excessive, pressure to stock being sanded/grinded.
10. Keep the work square on the belt.
11. Hold the work so it will not run off the edge of the belt or get caught under the belt.
12. Use the entire width of the belt.
13. Hold sharp pointed metal so it will not jam and catch into the belt.
14. Turn off the grinder before leaving it and make sure it has completely stopped.
15. Keep a pail of water nearby in which to cool metal.
16. When grinding long material, support one end on a solid rest or have someone help by holding it.
17. Identify any pinch points prior to using the tool.
18. Identify the location of the tool shut-off for use in an emergency situation.

GENERAL SAFETY PRACTICES

1. Wear approved eye protection, hearing protection, and proper clothing. Tie up loose hair and remove loose jewelry.
2. Do not operate the machine without the instructor's permission, or without instructor supervision.
3. Wear a dust mask or respirator when using the metal belt grinder for prolonged periods of time.
4. Do not talk to anyone while operating the metal belt grinder.
5. Be sure the main power switch is off or the cord is disconnected before making any adjustments or repairs on the grinder.
6. Keep fingers away from the belt.
7. Do not wear gloves; they may get caught in the belt.

8. Use vise grips to hold small pieces.

9. Do not leave the grinder until it has been shut off and has stopped running.

10. Clean off chips with a brush.

11. Do not rub eyes before washing any chips and grit off hands. Small pieces of metal may get into eyes and cause serious injury.

12. Do not use the grinder if the light is not adequate.

13. Keep the floor around the grinder clear of scraps, dust and other material.

14. Keep flammable materials, such as gasoline, away from the metal belt grinder.

COMPLETION QUESTIONS:

1. The belt should run on the ____________________________ of the pulleys.

2. The ____________________________ is spliced for operation in only one direction.

3. The table should clear the belt by ____________________________ inches.

4. If the belt is too loose, it will ____________________ ____________________ in front of the work.

5. The belt will run toward the ____________________________ side of the pulleys.

6. If the belt is too tight, the ____________________________ will wear excessively.

7. ____________________________ the motor before applying work to the belt.

8. The ____________________________ pulley can be moved to adjust the tightness of the belt.

9. If the belt runs in the wrong direction, it will ____________________________.

10. The work should not ____________________ ____________________ the edge of the belt.

METAL CUT OFF SAW
(ABRASIVE WHEEL SAW OR CHOP SAW)

PART IDENTIFICATION

Identify the circled parts on the metal cut off saw (abrasive wheel saw or chop saw) illustrated below.

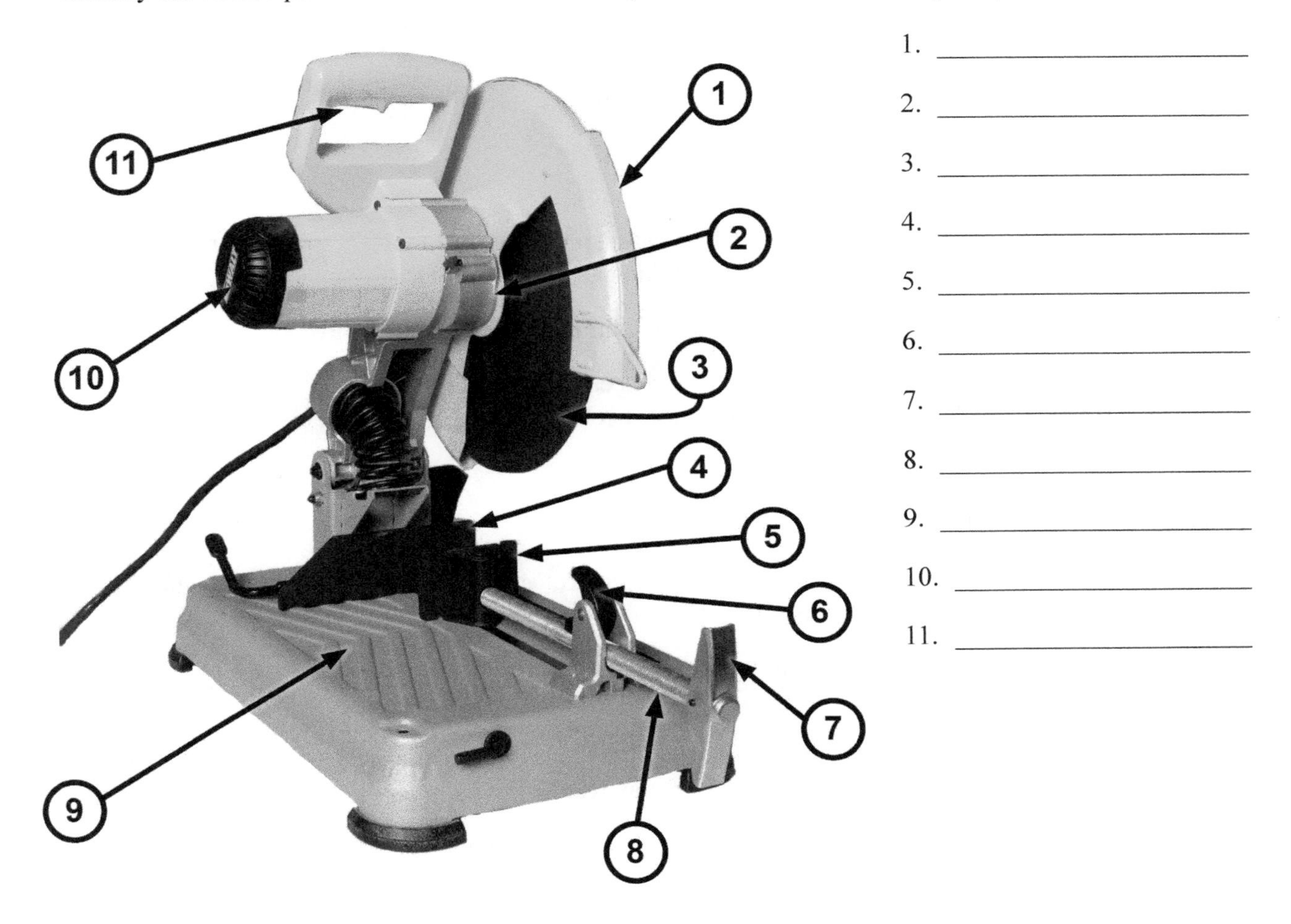

1. ____________________
2. ____________________
3. ____________________
4. ____________________
5. ____________________
6. ____________________
7. ____________________
8. ____________________
9. ____________________
10. ____________________
11. ____________________

SAFE OPERATIONAL PROCEDURES

1. Use a square to make sure the fence is truly square with the abrasive wheel or blade before clamping material into it. Use the wrench that comes with the machine to square it up if it has been knocked out of squareness.
2. Measure your material and mark it with a line. Line the blade up on the waste side of the line. Make sure the stock is tightly clamped in place before starting your cut. If it isn't, the blade will grab it and roll it around, or kick back into the blade or wheel, causing it to explode.
   a. When cutting short length stock, another piece of the same material should be clamped to the left side of the clamp to even out the clamping force.

b. When cutting long stock, it should be properly supported.

c. When cutting something small height in diameter or thickness, clamp another piece of the same material on top of the clamp to even out the clamping force.

d. Flat bar should be clamped vertically in the clamps.

e. Angle iron should be clamped with two edges touching the fence, and the point touching the clamp.

f. Extremely thin metal and material that is too wide should not be cut on the chop saw.

3. Never place your body or fingers in line with the blade or wheel while cutting.
4. Squeeze the trigger and allow the motor to reach full speed before contacting the material.
5. Do not force cutting. Always start the cut gently. Do not bump or bang an abrasive wheel or blade down on the material to start a cut.
6. If the blade or wheel binds or stops rotating, or the motor sounds like it is straining, release the switch immediately to reduce the risk of damage to the machine.
7. Make short plunging cuts up and down multiple times to allow the blade to cool during a cut.
8. Release the trigger after a cut is complete, and keep the blade or wheel away from your body until it has stopped. Be aware that blades and wheels may coast after the machine is turned off.

GENERAL SAFETY PRACTICES

1. Wear approved eye protection, face shield, hearing protection, welding coat or leather apron and proper clothing. Tie up loose hair and remove loose jewelry.
2. Do not operate the machine without the instructor's permission, or without instructor supervision.
3. Never force a blade or wheel onto an arbor or alter the size of an arbor. Do not use a blade or wheel that does not fit the arbor, as vibration may result.
4. Make sure the speed marked on wheel is at least as high as the no load RPM on the tool.
5. Set the machine securely on a flat, level surface. Before installing a wheel or blade, always check it for damage. With the machine unplugged and the machine head all the way down, manually spin the blade or wheel to check for wheel clearance and alignment. The blade or wheel should rotate freely and not contact the table. Replace cracked abrasive wheels or damaged blades immediately.
6. Be sure all guards are in place and working properly before each use.
7. Store blades and wheels with care. Do not drop them or subject them to excessive heat, cold or humidity.
8. When cutting metal, sparks or hot fragments could cause fires or burns. Never touch a work piece until it cools and take care when handling so you do not cut yourself on the burrs made by the machine.

9. If metal chips or filings build up in the saw, turn it off and clean it with a brush. Never use your hand. Metal slivers really hurt.

COMPLETION QUESTIONS

1. When cutting long stock, it should be properly ________________.
2. Flat bar should be clamped ________________ in the clamps.
3. Never place your ________________ or ________________ in line with the blade or wheel while cutting.
4. Do not ________________ cutting. Always start the cut gently.
5. Release the ________________ after a cut is complete, and keep the blade or wheel away from your body until it has stopped.
6. Do not use a blade or wheel that does not fit the ________________, as vibration may result.
7. Make sure the speed marked on wheel is at least as high as the no load ________________ on the tool.
8. Before installing a wheel, always check it for ________________.
9. Replace ________________ abrasive wheels immediately.
10. Store blades and wheels with care. Do not drop them or subject them to ________________ ________________, cold or humidity.

METAL LATHE

SAFE OPERATIONAL PROCEDURES

1. Facing off round material:
   a. Select a piece of round stock that fits in the chuck.
   b. Mount material in the lathe using a three-jaw universal chuck with about 1-1/8" of material sticking out of the chuck.
   c. Select a correctly sharpened right-handed general purpose turning tool.
   d. Mount a left-handed tool holder in the tool post, insert the turning tool in the tool holder, and tighten the set screws with the cutting edge at the same level as the center of the material.
   e. Rotate spindle by hand to ensure its free movement.
   f. Set spindle speed at 600 RPM.
   g. Start machine.
   h. Face off end of material by feeding the tool into the center of the material with the carriage handwheel and then slowly moving the tool back toward the operator using the cross slide handwheel.
2. Drilling center holes with a lathe chuck:
   a. Select a universal three-jaw chuck. Protect ways with a board.
   b. Wipe the cone surfaces clean with a rag and mount the chuck on the headstock.
   c. Securely fasten a short shaft in the chuck with 1" to 2" extending to the right of the jaws.
   d. Face off the end of the shaft as described above.
   e. Mount the drill chuck in the tailstock after wiping the tapered surfaces clean.
   f. Securely mount a combination center drill and countersink in the drill chuck. Tighten chuck in all three tightening positions with proper chuck wrench.
   g. Move the combination center drill and countersink close to the faced off material by sliding the tailstock. Clamp down the tailstock.
   h. Set the spindle speed at 150 RPM.
   i. Start the motor.
   j. Use the tailstock handwheel to slowly feed the combination center drill and countersink into the material.
   k. Stop drilling when the outside diameter of the countersink hole is 3/16" if the diameter of the countersink body is 1/4".
   l. After the center holes have been properly drilled in both ends of the material, it is ready to be mounted between centers.
3. Mounting the work between centers:
   a. Lubricate center holes with a high-pressure lubricant.

b. After wiping clean the tapered surfaces, wedge a dead center in the headstock and a live or dead center in the tailstock.
c. Mount a face plate on the headstock.
d. Slide a lathe dog over the material.
e. Hold the material between the centers with one hand and move the tailstock up near the end of the material.
f. Clamp the tailstock to the bed.
g. Snug the centers into the center holes.
h. Place the lathe dog into a notch in the face plate.
i. Tighten the set screw in the lathe dog.
j. The lathe dog should move freely back and forth in the face plate notch.
k. While moving the dog back and forth, tighten the centers by rotating the tailstock handwheel until a slight tension is felt.
l. Clamp tailstock spindle.

GENERAL SAFETY PRACTICES

1. Wear approved eye protection, hearing protection, and proper clothing. Tie up loose hair and remove loose jewelry.
2. Do not operate the machine without the instructor's permission, or without instructor supervision.
3. Remove all wrenches, oil cans, and other materials from the work area before starting the machine.
4. Be sure the chuck is tightly mounted on the spindle.
5. Be sure the material is tightly mounted in the chuck.
6. Do not leave a chuck wrench in the chuck at any time.
7. The operator should always start and stop the machine.
8. Never reach across the work while the machine is running.
9. Use a brush or hook to remove all chips.
10. Stop the machine before making any adjustments.
11. Stop the machine for all measurements.
12. Stop the feed before the tool reaches the jaws of the chuck.
13. Identify the location of the tool shut off for use in an emergency situation.

COMPLETION QUESTIONS

1. A ____________________________ ____________________________ tool holder works best for facing up round material.
2. When facing round material, it should extend ____________________________ inches to the right of the chuck jaws.
3. Before starting the machine, the chuck should be ____________________________ by hand to ensure its free movement.
4. A ____________________________ hand general purpose turning tool should be used to face off round material.
5. The spindle should rotate at about ____________________________ RPM when facing off round material.
6. The ________________________ ________________________ handwheel and the ________________________ handwheel should be used to feed the turning tool into the work.
7. The ____________________________ post holds the tool holder in place.
8. The ____________________________ surfaces should be wiped clean before the drill chuck is mounted.
9. The spindle speed should be about ____________________________ RPM for drilling center holes.
10. Dead centers should be lubricated with a ____________________________ ____________________________ ____________________________.

MILLING MACHINE

PART IDENTIFICATION

Identify the circled parts on the milling machine illustrated below.

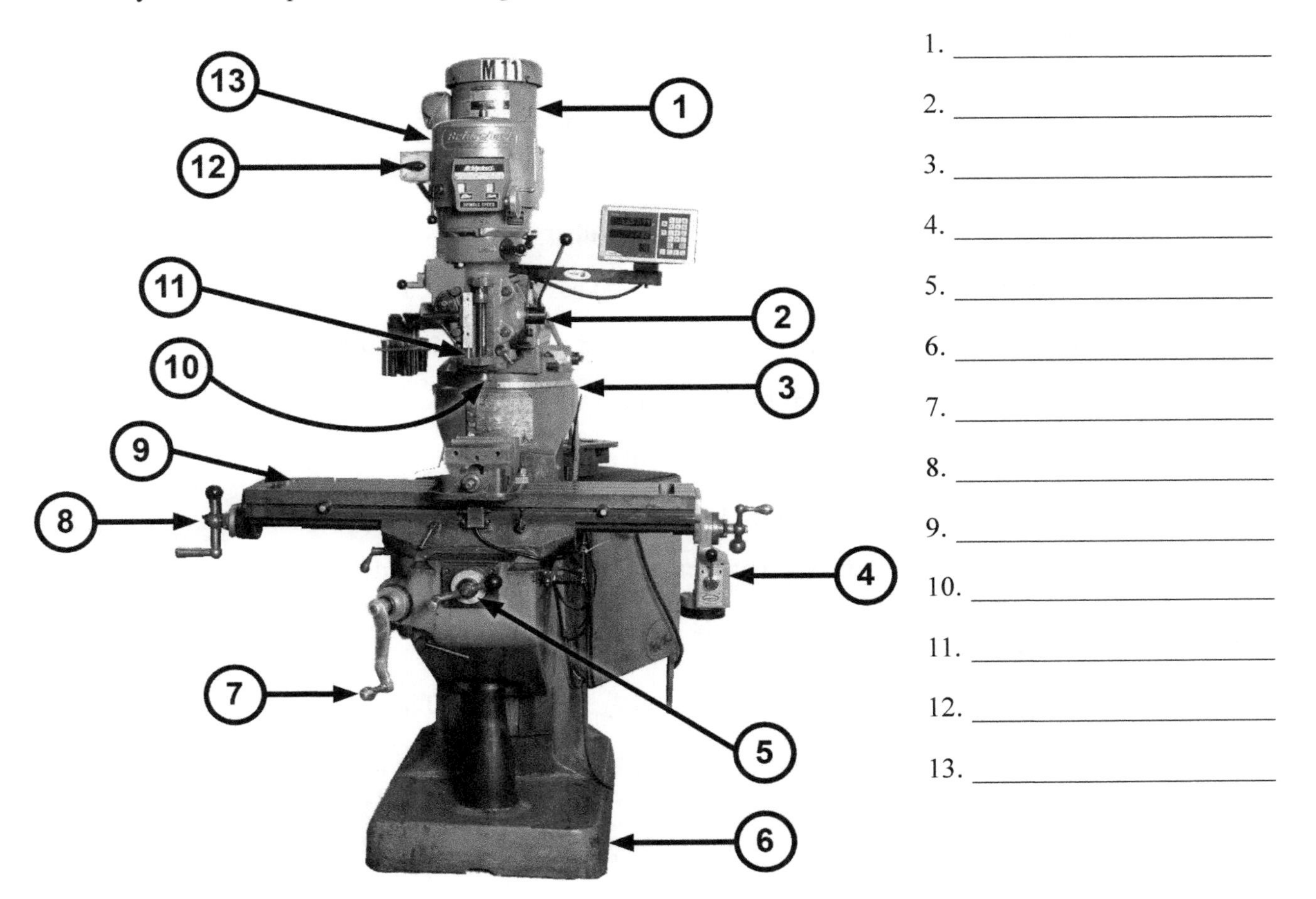

1. ______________________
2. ______________________
3. ______________________
4. ______________________
5. ______________________
6. ______________________
7. ______________________
8. ______________________
9. ______________________
10. ______________________
11. ______________________
12. ______________________
13. ______________________

SAFE OPERATIONAL PROCEDURES

1. Do not place anything on the milling machine table other than material and clamping fixtures.
2. When setting up a job, move the table with the material as far as possible from the cutter.
3. Your material must be properly clamped and secure when working with the mill.
4. Remove the collet tightening wrench immediately after using it.
5. Make sure the cutter is rotating in the proper direction before cutting.
6. Never run the mill faster than the recommended cutting speed.
7. Make sure the milling machine is completely stopped and cutter is not revolving before placing hands near the table to take measurements or manipulate the material.

8. Make sure the power is disconnected before changing cutters.

GENERAL SAFETY PRACTICES

1. Wear approved eye protection, hearing protection, and proper clothing. Tie up loose hair and remove loose jewelry.
2. Do not operate the milling machine without the instructor's permission, or without instructor supervision.
3. Never wear gloves when operating the milling machine.
4. Always use cutters that are sharp and in good condition.
5. Milling machine movements can be controlled by either hand feed or a power feed.
6. Always use cutting oil to keep the material cool and have the related SDS on hand.
7. Don't take too heavy of a cut or use to fast of a speed, this could fracture the cutter.
8. No food or beverages around the milling machine.
9. Keep your milling work area clean and well-lit at all times.
10. Always use a brush to clear metal chips from the work on the milling machine.
11. Make sure the milling machine is disconnected from electricity when doing maintenance.

COMPLETION QUESTIONS

1. Always ensure that _______________ are correctly sharpened and in good condition.
2. When setting up a job, move the _______________ with the material as far as possible from the cutter.
3. Never place hands near a _______________ cutter.
4. Always use a suitable _______________ to clear chips from your working area.
5. Never take an excessively _______________ cut or feed, this could fracture the cutter.
6. No _______________ or _______________ around the milling machine.
7. Work must be _______________ securely in a vise or on to the table.
8. Always check the feed and _______________ of the mill cutter to prevent injury of the operator.
9. Remove the collet tightening _______________ immediately after using it.
10. Make sure the power is _______________ before changing cutters.

MORTISING MACHINE

PART IDENTIFICATION

Identify the circled parts on the mortising machine illustrated below.

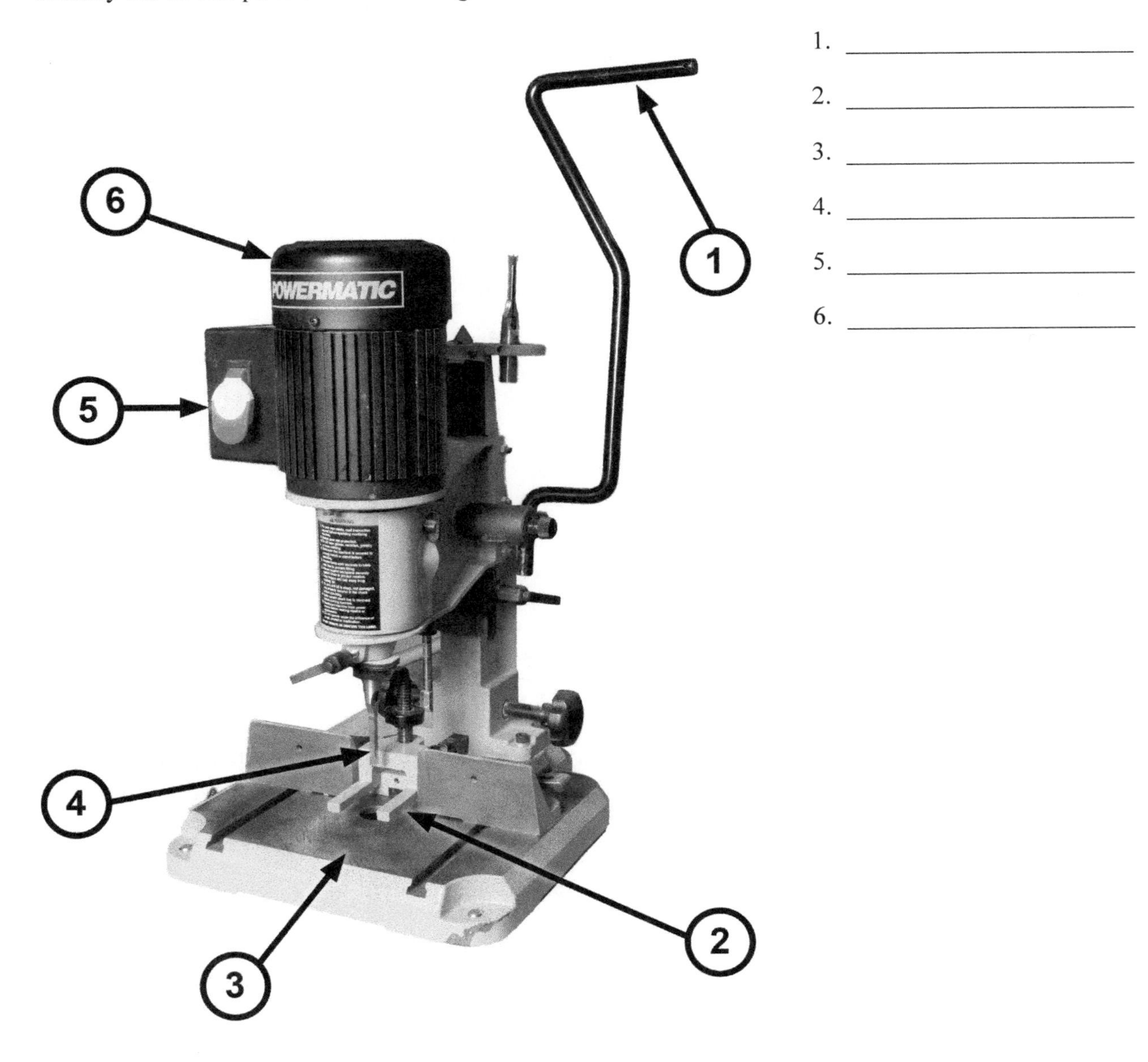

1. ______________________
2. ______________________
3. ______________________
4. ______________________
5. ______________________
6. ______________________

SAFE OPERATIONAL PROCEDURES

1. Read operators manual and risk assessment.
2. Keep work area clean and tidy. Ensure no slip/trip hazards are present in the workspace or walkway.
3. Check the condition of the auger and chisel. They should be kept sharp at all times.

4. Check the operation of all fences, holding clamps and machine guards.
5. Check all adjustments and settings carefully before commencing task.
6. Check for alignment, binding and breakage of moving parts.
7. Cut centered on the chisel.
8. When positioned correctly, lock the fence in place by tightening the levers.
9. Never attempt to drill material that does not lie flat on the table.
10. Irregular shaped work should be securely held in a jig, vice or clamp.
11. Set the work piece on the table so that the chip clearing slot of the chisel faces the direction you will move the work piece to make your next cut.
12. Tighten the table clamp to hold the work piece securely against the fence.
13. Lower the work piece hold-down until slightly above the work piece and tighten.
14. Make sure the two side plates of the chuck cover are closed.
15. Make Sure the chisel and bit are properly installed and secured.
16. Turn the Power "ON". Check that the machine runs without vibration or shaking. Visually check that the bit does not wobble.
17. Keep clear of moving machine parts.
18. Proceed with cut. Lower the chisel into the work piece using a steady even pressure until it bottoms out on the depth stop.
19. Raise the chisel. Re-position work piece to continue the cut.
20. Repeat cuts with a slight overlap until mortise is complete.
21. The workpiece should be moved so that the chisel is releasing chips into the already cut part of the workpiece.
22. Allow the auger bit to cut at its own speed without applying excessive pressure.
23. Keep hands clear of the auger and chisel when the machine is running.
24. Do not remove waste material from the table while the machine is running.
25. Turn OFF the power source when finished. Never leave the machine running unattended.

GENERAL SAFETY PRACTICES

1. Wear approved eye protection, hearing protection, and proper clothing. Tie up loose hair and remove loose jewelry.
2. Do not operate the machine without the instructor's permission, or without instructor supervision.
3. The working area should be well lit, clean and free of debris.
4. Inspect machine for broken or missing parts.

5. Inspect cords for frays or cuts, report the problem for repairs if cord is damaged.
6. Inspect lumber for foreign materials—(nails, screws, etc.).
7. Layout and mark work pieces for the mortises to be cut. Transfer a reference line showing the desired depth of cut onto the front face of the work piece.
8. Always select the correct auger and chisel set for the mortise to be drilled and install them in the chuck.
9. Use the correct key and tighten the chuck uniformly when the auger bit is inserted.
10. Always remove the chuck key before starting the mortising machine.
11. Place the material on the table against the fence.
12. Set the depth stop as follows: lower the chisel to the bottom of your depth reference line then adjust the depth stop.
13. Before making adjustments and measurements or cleaning swarf accumulations, switch off and bring the machine to a complete stop.
14. Clean bench and work area & place all waste material in bin.
15. Leave the machine in a safe, clean and tidy state.

COMPLETION QUESTIONS

1. Always position the _______________ _______________ _______________ directly over workpiece to prevent workpiece from lifting during operation.
2. Always support workpiece securely against the _______________ to prevent rotation.
3. Never turn on the _______________ with the drill bit or chisel contacting the workpiece.
4. Adjust the _______________ stop to avoid drilling into the table.
5. The opening on the side of the chisel should always be to the _______________ or _______________, never to the front or rear.
6. Always remove the _______________ _______________ before starting the mortising machine.
7. Inspect lumber for _______________ materials.
8. Leave the machine in a _______________, _______________ and tidy state.
9. Inspect machine for _______________ or missing parts.
10. Do not remove waste material from the table while the machine is _______________.

OSCILLATING SPINDLE SANDER

PART IDENTIFICATION

Identify the circled parts on the oscillating spindle sander illustrated below.

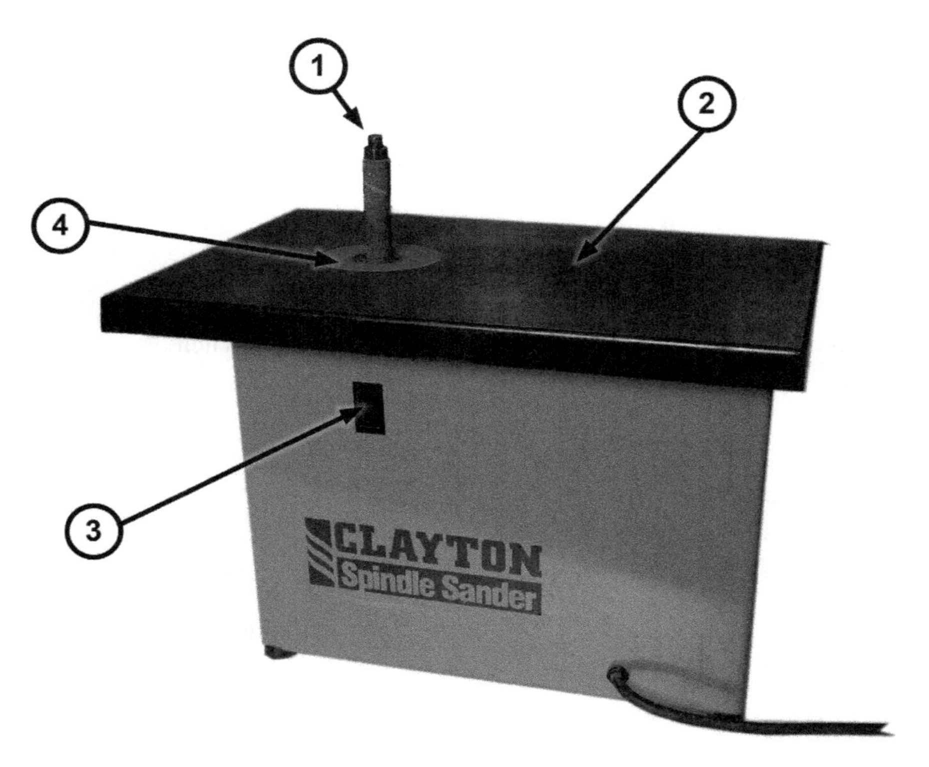

1. ____________________
2. ____________________
3. ____________________
4. ____________________

SAFE OPERATIONAL PROCEDURES

1. The oscillating spindle sander can be used to sand concave edges. Never sand flat or convex surfaces or you will develop divots in the edge.
2. Select the coarseness of the abrasive according to the more common sanding jobs to be completed. Keep the sandpaper clean and in good condition.
3. Use the appropriate sized table insert to keep the distance between the table and abrasive as small as possible.
4. For the smoothest results, install the largest diameter spindle that will fit the contour you are sanding. Do not sand a large diameter contour with a small diameter spindle.
5. Do not sand material that is too small to be safely supported.
6. Do not operate the sander with a damaged spindle or sanding sleeve.
7. Always feed work against the direction of spindle rotation.

8. With the sander off, determine a grip on the material that will not allow your fingers to contact the spindle. Make sure you can sand the material left and right of center while always sanding against the rotation of the spindle.
9. Turn the motor on, allow the motor to gain full speed, and then move the material carefully into the spindle with enough pressure to keep the sandpaper cutting.
10. Hold the material down against the table firmly at all times.
11. When the material is smooth or to the desired sanding line, remove it from the table and turn off the motor.
12. Do not leave the safety zone area until the spindle completely stopped.

GENERAL SAFETY PRACTICES

1. Wear approved eye protection, hearing protection, and proper clothing. Tie up loose hair and remove loose jewelry.
2. Do not operate the machine without the instructor's permission, or without instructor supervision.
3. Never wear gloves while operating the sander.
4. With the machine off, inspect the sandpaper to make sure it is clean and free from damage. Never operate the machine if the sandpaper is loose, torn, or filled with sanding dust.
5. Make all adjustments with the motor off and the spindle completely stopped.
6. Allow the machine to reach full operating speed before beginning to sand.
7. Maintain a balanced stance and keep your body under control at all times. Do not overreach.
8. Do not allow hands or fingers to get near or touch the moving abrasive/sandpaper. On small or thin material use a jig to keep the hands from contacting the abrasive.
9. Be sure the table is locked in position before placing material on it and that the table insert is within an 1/8" or less of the abrasive/sandpaper.
10. Do not overload the motor with excessive pressure; instead use light pressure and keep your workpiece moving to avoid burning.
11. The spindle sander is designed to smooth edges or end grain of material and not for cutting excessive amounts from edges or ends of boards.
12. Long periods of continuous sanding will overheat the spindle, causing the material to be discolored and possibly damaging the machine. Cut to within ⅛" of your finished line before using the sander.
13. Never clean the table with power on. Never use your hands to clear sawdust and debris; use a brush.
14. Before leaving the machine, the sander should be turned off and the spindle should be completely stopped.

15. Sand the material in the opposite direction of the spindle's rotation.
16. Keep the machine guards in place at all times when the machine is in use. Remove the energy source before removing guards to perform maintenance.

COMPLETION QUESTIONS

1. Never clean the table with power on. Never use your _______________ to clear sawdust and debris; use a brush.
2. You can use the _______________ _______________ sander to sand concave shapes.
3. The _______________ can be tilted for sanding bevel edges.
4. The oscillating spindle sander can sand _______________ of material.
5. Sanding should be done moving in a _______________ direction to the spindle.
6. The edge of the tilting table should never be more than _______________ inch from the disc.
7. Before leaving the machine, the sander should be turned off and the spindle should be completely _______________.
8. Holding your material in one spot or applying pressure will cause your material to _______________.
9. A _______________ could be used to check the angle between the table top and sanding spindle.
10. Allow the sander to obtain full _______________ before touching your material to the sandpaper.

PANEL SAW

PART IDENTIFICATION

Identify the circled parts on the panel saw illustrated below.

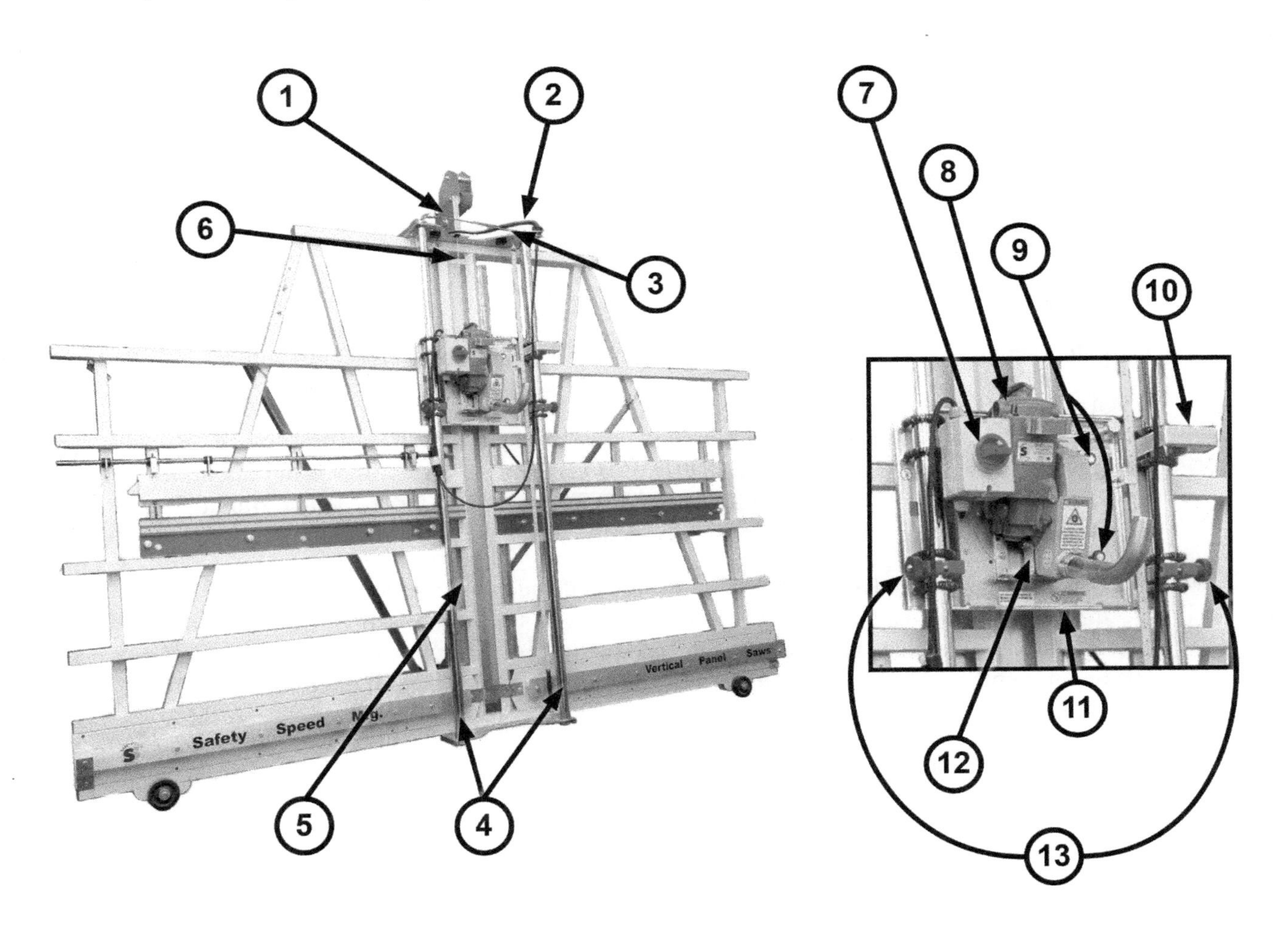

1. ______________________________

2. ______________________________

3. ______________________________

4. ______________________________

5. ______________________________

6. ______________________________

7. ______________________________

8. ______________________________

9. ______________________________

10. ______________________________

11. ______________________________

12. ______________________________

13. ______________________________

SAFE OPERATIONAL PROCEDURES

1. Panel saws usually have predefined minimum and maximum workpiece sizes. These are based upon the physical configuration of the panel saw horizontal and vertical tracks and the directionality of the saw configuration. Most accept standard 4 x 8-foot sheet goods and can make full length horizontal or vertical cuts in them.
2. Ensure that the area around the saw is safe for workpiece travel and that any assistants are properly briefed on sawing operations.
3. If using a stop, set it at this time. Bring saw to a complete stop every time you adjust stop.
4. Check direction of feed when ripping material.
5. Never rip with larger material at top side of blade.
6. Open blast gate and turn dust collection on, hook up hose.
7. Turn on saw—allow for saw to accelerate to full speed prior to the cutting operation.
8. When crosscutting, with a firm grip slowly bring saw down and engage workpiece at a constant speed. Continue cutting in a smooth downward motion till saw passes thru workpiece. ALWAYS CROSSCUT FROM TOP TO BOTTOM. Turn off saw at bottom of stroke and allow blade to come to a COMPLETE STOP before slowly—with a firm hold—allowing the saw to retract to the top retracted position.
9. When ripping, smoothly feed workpiece into sawblade and continue feeding at a constant speed. Special care must be taken as the operator and/or assistant transition from pushing the material into the blade to pulling the material thru from the opposite side of the saw. The upper cutoff will need to be supported as the cut proceeds so that it does not pinch the sawblade. DO NOT USE PUSHSTICKS
10. Turn off saw and allow blade to come to a COMPLETE stop. Remove workpieces from saw.
11. Do not force saw when cutting.
12. Shut blast gate and turn off dust collection unless you are proceeding to another machine or someone else is using it.
13. Clean up around saw.
14. Tear down all setups made. Setups may be left with approval of supervisor.
15. Workpieces for rip cutting generally need to be supported on a minimum of 4 bottom support rollers.
16. Workpieces for crosscutting generally need to be wider than the carriage of the saw for proper clamping and support.

GENERAL SAFETY PRACTICES

1. Wear approved eye protection, hearing protection, and proper clothing. Tie up loose hair and remove loose jewelry.

2. Do not operate the machine without the instructor's permission, or without instructor supervision.
3. Never saw without piece against supports.
4. Never adjust saw or setup while saw is running.
5. Always follow 4-inch rule (always keep hands 4 inches away from blade).
6. Don't reach across the blade.
7. Never remove guard unless authorized by instructor.
8. Any operation other than a standard crosscut or rip cut must be approved by instructor. This includes any blade changes.
9. Never run materials containing nails, screws or other metallic objects.
10. Never cut small pieces in panel saw (less than 29"x29").
11. Never cut more than one piece at a time.
12. If machine is malfunctioning stop immediately and report to instructor.

COMPLETION QUESTIONS

1. Ensure that the area around the saw is _______________ for workpiece travel
2. Never rip with larger material at _______________ side of blade.
3. Turn on saw—allow for saw to accelerate to _______________ speed prior to the cutting operation.
4. Always crosscut from _______________ to bottom.
5. Workpieces for rip cutting generally need to be supported on a minimum of _______________ bottom support rollers.
6. Never cut small pieces in panel saw less than _______________ X _______________.
7. If machine is malfunctioning stop immediately and _______________ to instructor.
8. Never saw without piece against _______________.
9. Never run materials containing nails, screws or other _______________ object.
10. Turn off saw and allow _______________ to come to a COMPLETE stop.

PARTS WASHER/SOLVENT TANK

PART IDENTIFICATION

Identify the circled parts on the parts washer/solvent tank illustrated below.

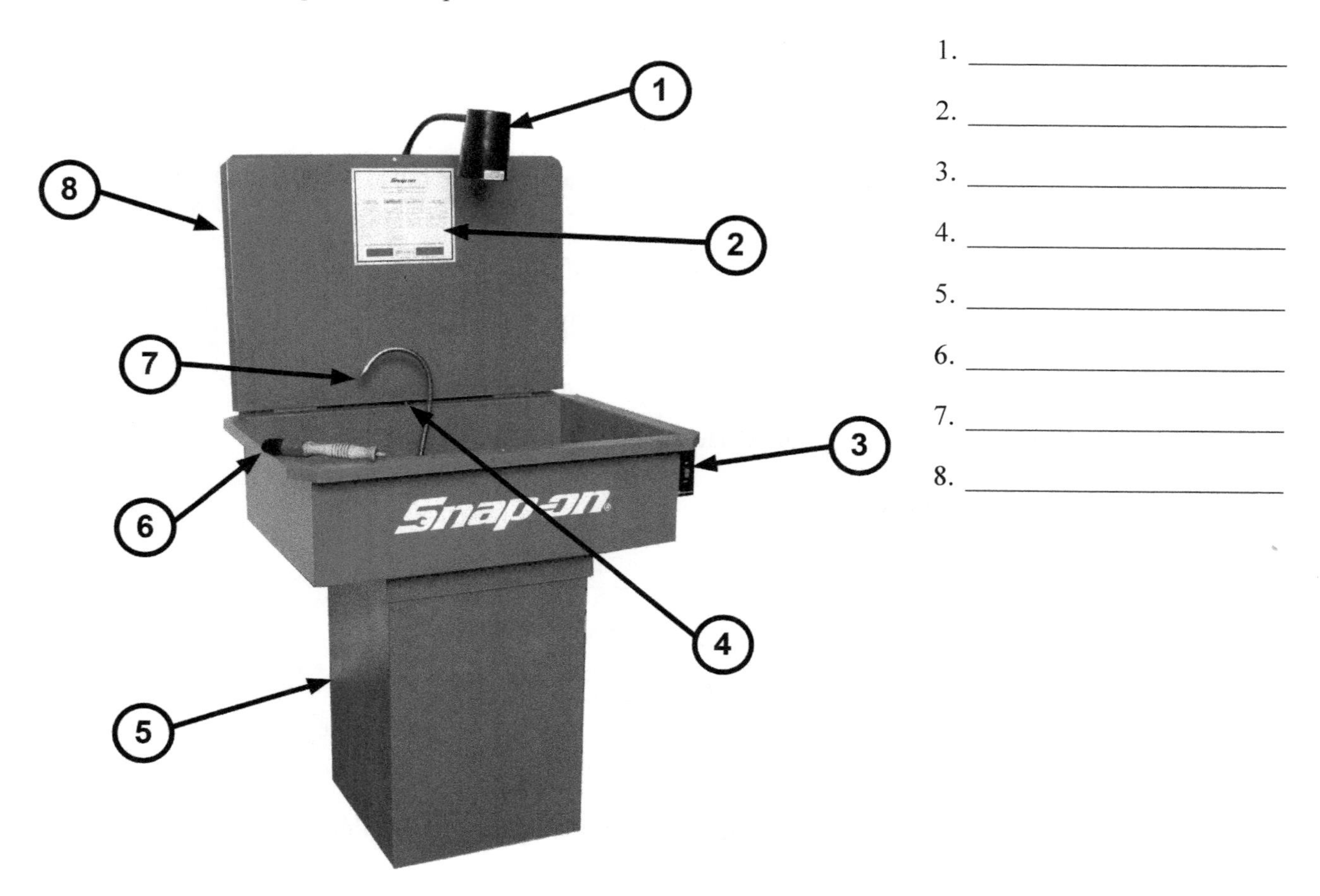

SAFE OPERATIONAL PROCEDURES

1. Perform a pre-operational inspection: Locate & ensure you are familiar with all washer operations & controls.
2. Ensure that a parts washer that uses combustible or flammable solvents is equipped with a self-closing lid if a fire occurs, it is usually a fusible link (holding device that will melt and drop the lid in the event of a fire).
3. Ensure there is adequate ventilation and ensure that no ignition sources are within 25 feet of the parts washer.
4. Perform operational safety checks: Be aware of other people in the area of the parts washer, make sure the area is clear before starting the washer and keep the lid closed when not in use.

5. Ending operations & clean up: Turn off the washer, close the lid, wash hands with soap & water after every use, clean up all spills immediately, do not leave dirty or soiled rags lying around and leave the work area in a safe, clean, and neat condition.

6. Potential hazards: Electric shock, degreasing fluid can affect the skin sharp edges & burrs on parts being cleaned and strains & sprains from handling parts.

7. Consult the Safety Data Sheets (SDS) for specific technical data and precautionary measures concerning the use of the parts washer.

GENERAL SAFETY PRACTICES

1. Wear approved eye protection (goggles), hearing protection, gloves, and proper clothing. Tie up loose hair and remove loose jewelry.

2. Do not operate the machine without the instructor's permission, or without instructor supervision.

3. Use proper lifting techniques or ask for help when placing items in wash tank.

4. Use proper cleaning solutions. Never use highly volatile flammable solvents such as gasoline, toluene, diesel fuel, methyl ethyl ketone (MEK), or 1,1,1-trichloroethane. Read and understand all information concerning cleaning solvents before using.

5. Do not use or store near open flames, pilot lights in stoves or heaters, or any other ignition source. Solvents are highly flammable.

6. Materials used when cleaning and debris removed by cleaning may be harmful or fatal if inhaled or swallowed.

7. Plug into GFCI protected outlet only and use in a well-ventilated area.

8. The use of accessories or attachments not recommended by the manufacturer may result in a risk of injury to persons.

9. Stay alert, watch what you are doing and use common sense when operating a parts washer/ solvent tank.

10. Keep the work area clean, tidy, well swept/washed, and well lit; floors should be level and have a non-slip surface.

11. Do not remove the lid and fusible link; make sure that they are in position and in good working condition before operating.

12. When brushing parts in solvent, use a nylon or brass bristle brush to avoid sparks.

COMPLETION QUESTIONS

1. PPE (Personal Protective Equipment) includes _______________ eye protection.

2. Be aware of other _______________ in your work area when using the parts washer.

3. Hand protection includes wearing ______________ that are compatible with the solvent used.

4. Ensure that there is adequate _______________ when using the parts washer.
5. Keep the parts washer _______________ closed when not being used.
6. Do not operate the machine without _______________ supervision.
7. Check workspaces and walkways to ensure no _______________ hazards are present.
8. Always use correct _______________ technique when putting parts in wash tank.
9. Do not operate parts washer if the lid is _______________ or faulty.
10. Switch off the parts washer machine when it is not in _______________.

PLANER/SURFACER

PART IDENTIFICATION

Identify the numbered parts of the planer/surfacer illustrated below.

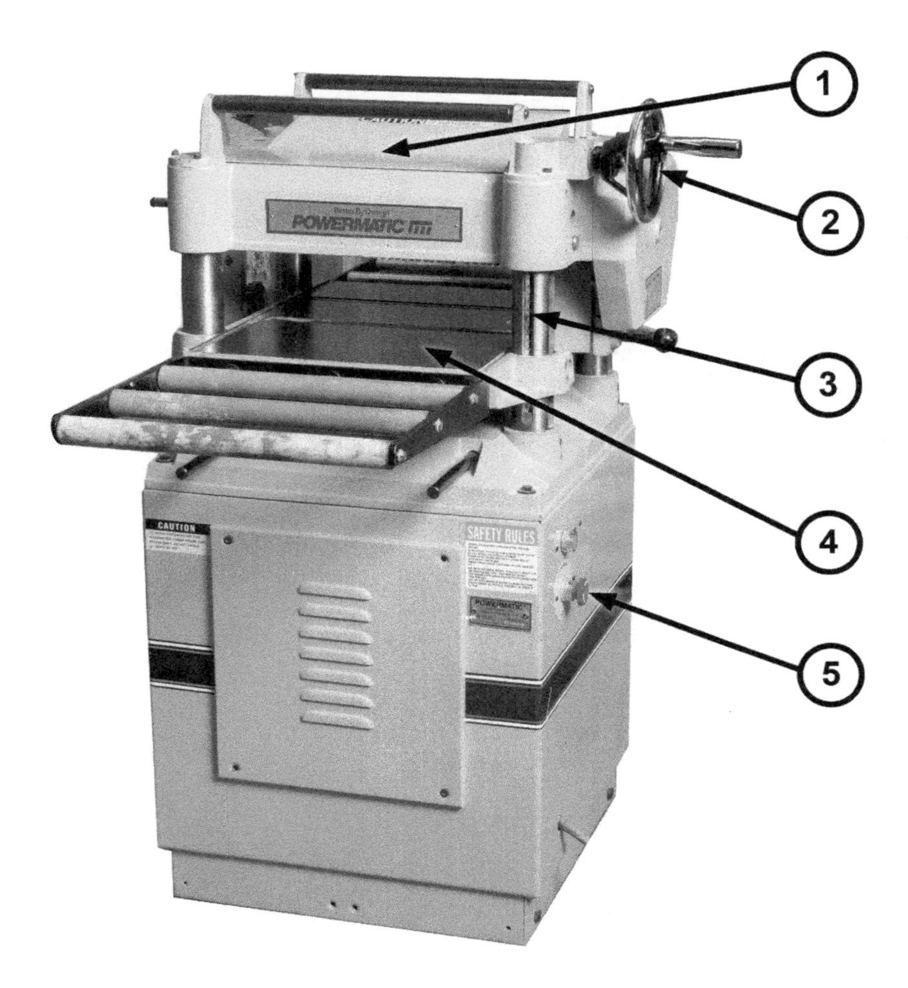

1. ____________________
2. ____________________
3. ____________________
4. ____________________
5. ____________________

SAFE OPERATIONAL PROCEDURES

1. The planer/surfacer is designed to machine material to exact thickness. It is equipped with a cutterhead, and it is similar to a jointer except it cuts the material from the top.
2. The machine is sized by the width of the cutterhead with common sizes being 12”, 18”, and 24”.
3. It is equipped with infeed and outfeed rollers to automatically feed the material through the machine.
4. The depth of cut is determined by the location of the lower table bed in respect to the cutterhead, based on the starting thickness of the material.
5. To plane a board to exact thickness, measure the material at its thickest point.

6. Set the planer/surfacer for the thickness of the material minus the depth of cut using the height adjustment handwheel. The recommended depth of cut for one pass is no more than 1/16 of an inch for hardwoods and no more than 1/8 of an inch for softwoods.
7. Position the material so the knives will cut with the grain, with the first or true surface being down. Never plane or surface finished material, or material containing nails or other foreign material.
8. Turn on the machine and allow the motor to gain full speed.
9. Feed the material in at right angles to the cutterhead.
10. Do not force the material through the machine, but allow the feed rollers to pull the material through the machine.
11. If planing long material, get assistance from a helper or use a support stand.
12. Keep your fingers away from the pinch point area between the material and the infeed table.
13. Run the material through as many times as necessary to reduce to the desired thickness. Successive cuts should be taken off alternate faces. Make sure to plane with the grain.
14. Shut off machine and do not leave area until the machine has completely stopped.

GENERAL SAFETY PRACTICES

1. Wear approved eye protection, hearing protection, and proper clothing. Tie up loose hair and remove loose jewelry.
2. Do not operate the machine without the instructor's permission, or without instructor supervision.
3. Make sure knives are sharp and properly adjusted.
4. Make all machine adjustments before connecting to the energy source.
5. Keep floor and work area free of chips, wood scraps, and other materials.
6. Make sure motor is at full speed before feeding material into machine.
7. Always stand to one side when feeding or receiving material.
8. Do not place hands near feed rollers or knives.
9. If material gets stuck, shut machine off, wait for the machine to come to a full stop, raise the cutter head, and use a push stick to remove the material.
10. Remove loose knots, nails, or other defects before planing.
11. Do not plane material shorter than 12" because it will get caught between the infeed and outfeed rollers.
12. Do not plane material to a thickness of less than 3/8" without a backer board.
13. Never talk to others while operating the machine.

COMPLETION QUESTIONS

1. The planer/surfacer is sized by the ____________________ of the ____________________
2. There are ___________________________ and outfeed rollers in the machine that move the material through the machine.
3. Adjust the machine to the thickness of the material minus the depth of cut with the ________________________ ______________________ ________________________.
4. The maximum recommended cut per pass for softwoods is ________________________ of an inch.
5. Keep your ________________________ away from the pinch point area between the material and the infeed table.
6. Material shorter than ________________________ should never be planed.
7. If the material gets stuck in the planer, turn the power off, lower the bed using the handwheel and use a ________________________ to free the material.
8. The first run depth of cut is determined by the ________________________ part of the material.
9. The material should be fed into the machine at ________________________ angles to the cutterhead and so the knives are cutting the grain.
10. Material thinner than ________________________ of an inch should never be planed unless a ________________________ ________________________ is used.

POWER MITER SAW

PART IDENTIFICATION

Identify the numbered parts of the power miter saw illustrated below.

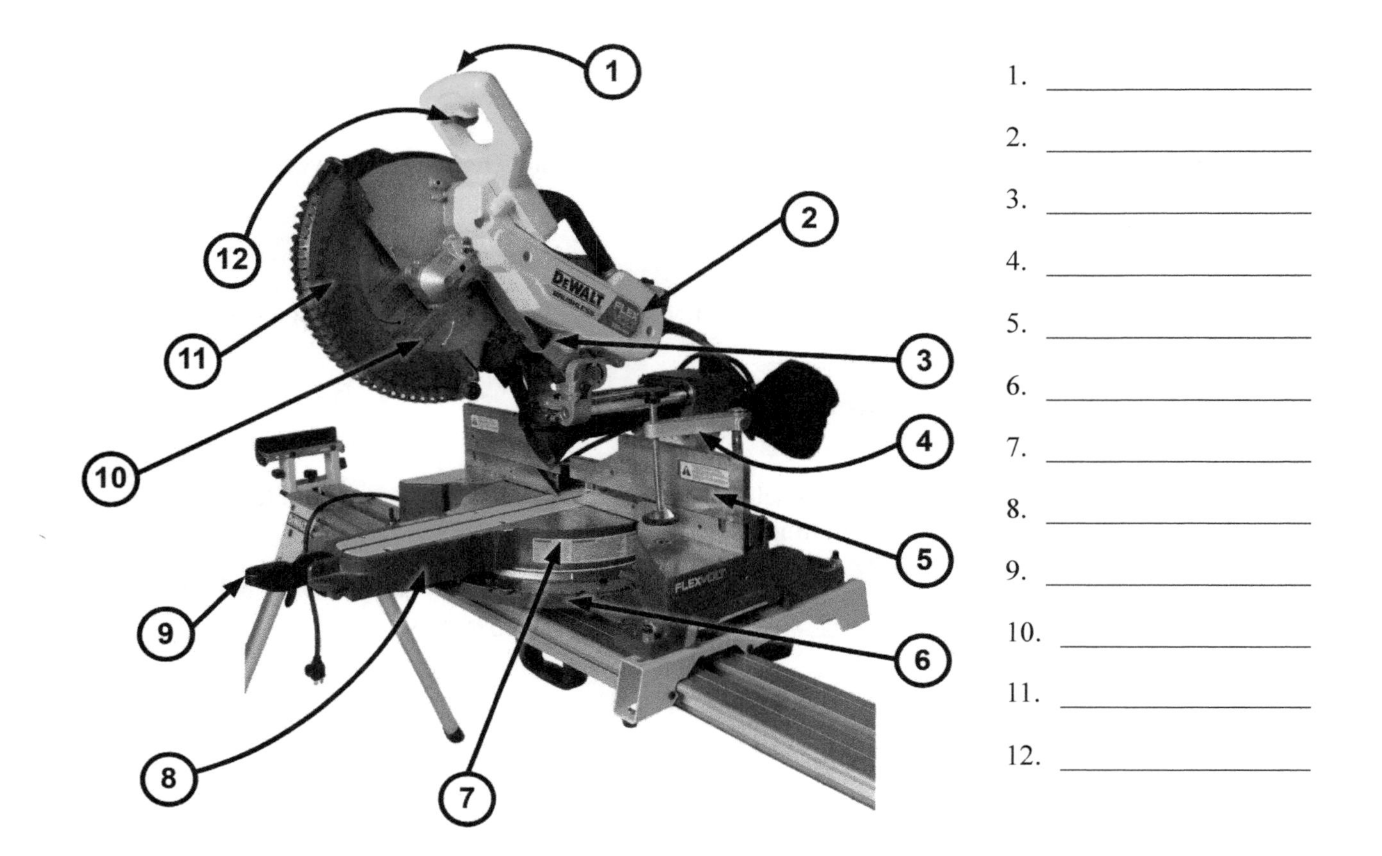

1. ____________________
2. ____________________
3. ____________________
4. ____________________
5. ____________________
6. ____________________
7. ____________________
8. ____________________
9. ____________________
10. ____________________
11. ____________________
12. ____________________

SAFE OPERATIONAL PROCEDURES

1. Study the operation, maintenance, and safety manual(s) for the specific saw to be operated.
2. Changing the saw blade:
   a. Disconnect saw from energy supply.
   b. Select a crosscut blade.
   c. Remove the saw guard, the arbor nut, and arbor collar. Remember the arbor has left-hand threads. Remove the blade.
   d. Noting the direction of rotation, place the blade on the arbor so the teeth point toward the operator and face downward.
   e. Place the outside collar on the arbor. Be sure the recessed face of both collars are against the saw blade.

f. Tighten the arbor nut using the wrench furnished with the saw.

g. Replace the guard. Make sure it moves freely before reconnecting energy supply.

3. Miter cuts:

a. Select a sharp crosscut blade.

b. The saw should be in the raised position when not in use.

c. Select the proper angle of cut by moving the spring-loaded miter arm. Most saws have a stop at 90 degrees and 45 degrees left and right. Lock the miter arm in position for the desired cut with the locking handle.

d. Place the material in the saw with the jointed edge against the fence.

e. Hold the material against the fence with one hand with the mark under the saw blade. Lower the blade down with your fingers off the trigger to align the blade with your mark. The blade should cut on the "scrap" side of your line.

f. Fingers or hands should never be within 4" of the path of the blade. Use a clamp to hold small pieces.

g. NEVER cross your arms while cutting with the miter saw. For example, don't hold your stock on the right side of the saw with your left hand and operate the saw with your right hand.

h. Start the motor before making contact with the material. Make sure saw is at full speed, then slowly lower the saw into the material with hand positioned on the saw handle and your other hand holding onto the stock.

i. After completing the cut, keep saw held down and allow the blade to come to a complete stop before removing material from the table. Then return the saw to the raised position.

GENERAL SAFETY PRACTICES

1. Wear approved eye protection, hearing protection, and proper clothing. Tie up loose hair and remove loose jewelry.

2. Do not operate the machine without the instructor's permission, or without instructor supervision.

3. Be sure the blade guard is in place and working properly.

4. Operate the miter saw only where adequate light is available.

5. Be sure the material is firmly supported. Do not attempt to hold the material away from the fence.

6. Always keep fingers more than 4" from the path of the blade.

7. Clean all scrap material and sawdust away from the work area before starting the saw.

8. Do not leave the work area until the saw blade has come to a complete stop.

9. When the job is completed and the saw is unplugged, clean the saw and work area.

COMPLETION QUESTIONS

1. The ____________________________ must be removed to change the blade.
2. The arbor nut has ____________________________ hand threads.
3. The teeth of the blade near the operator are pointed in a ____________________________ position.
4. After use, the spring returns the saw to the ____________________________ position.
5. When making a cut, the saw should be lowered ____________________________ into the material.
6. Allow the blade to attain ____________________ ____________________ before lowering it into the material.
7. The motor should be ____________________________ before removing material from the table or leaving the work area.
8. Fingers should never be closer than ____________________________ inches to the path of the blade.
9. A ____________________________ blade should be used for straight and miter cuts.
10. The ____________________________ ____________________________ is adjusted across the index or degree scale for making angle cuts.

RADIAL ARM SAW

PART IDENTIFICATION

Identify the numbered parts of the radial arm saw illustrated below.

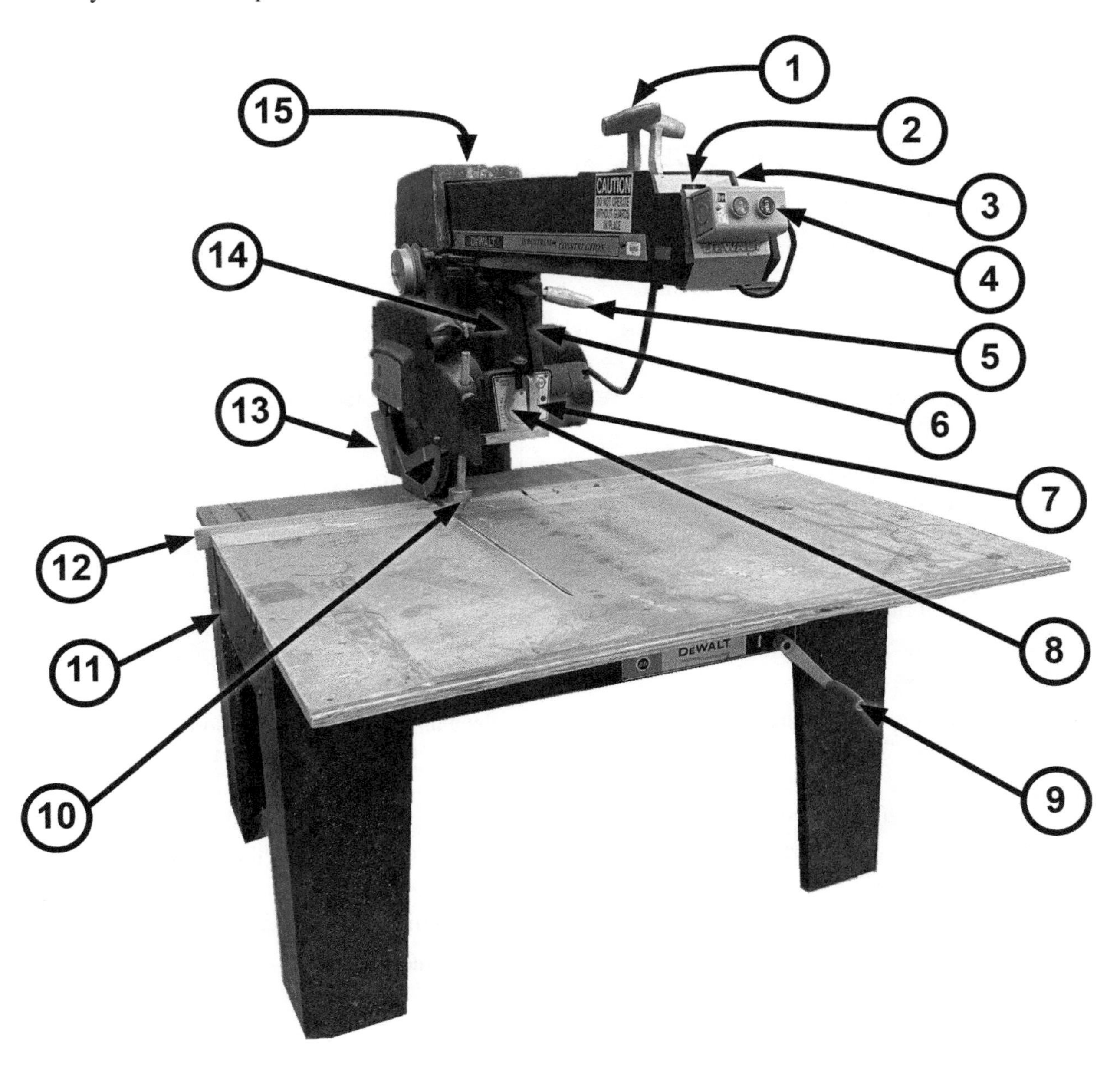

1. ______________________________

2. ______________________________

3. ______________________________

4. ______________________________

5. ______________________________

6. ______________________________

7. ______________________

8. ______________________

9. ______________________

10. ______________________

11. ______________________

12. ______________________

13. ______________________

14. ______________________

15. ______________________

SAFE OPERATIONAL PROCEDURES

1. Crosscutting:
   a. Select a crosscut or combination blade.
   b. Have all guards in place and make sure they are operating freely.
   c. Push the saw to the rear of the table. Tighten the rip lock to keep the saw from running forward when it is turned on.
   d. Adjust blade height by turning the elevating handle until the teeth just touch the table top.
   e. Adjust saw at a right angle to the fence.
   f. Place the material to be cut on the table with the straight, true & smooth edge tight against the fence and align the cut mark in line with the edge of the saw blade.
   g. Be sure the saw blade is not touching the material when the saw is turned on.
   h. Hold material with left hand and saw with right hand, standing slightly to the left of the line of the saw blade.
   i. Pull saw toward operator, using a controlled movement, feeding slowly enough that the saw does not grab.
   j. Return the saw to the rear of the table. Lock in place until ready to make another cut.
2. Ripping:
   a. Select ripping or combination blade.
   b. Turn saw parallel to the fence by releasing the swivel lock and turning the saw yoke. Lock in position with rip lock at the proper width of cut.
   c. Adjust safety guard and the anti-kickback device according to the operator's manual.
   d. Adjust the height of saw with elevating handle so that the teeth just touch the table.
   e. Feed material against the rotation of the saw blade.
   f. Use a push stick when working near the blade to keep hands away from the blade.
   g. Get an assistant to help with long material or use a roller support.
3. Miter cuts:
   a. Select a crosscut or combination blade.
   b. Set the motor yoke and the lock in the same position as for crosscutting. Release the arm clamp and the miter latch.

c. Swing the radial arm to the desired angle as indicated on the miter scale.

d. Re-engage the miter latch and tighten the arm clamp.

e. Make the cut in the same manner as described for crosscutting.

4. Bevel cuts:

   a. Select a crosscut or combination blade.

   b. Lock the radial arm and the motor yoke in the same position as for crosscutting.

   c. Raise the saw until the motor can be tilted to the desired bevel. Release the bevel clamp and the locating pin.

   d. Tilt the saw end of the motor downward to the desired bevel as indicated on the bevel scale.

   e. Re-engage the locating pin and tighten the bevel clamp.

   f. Make the cut in the same manner as described for crosscutting.

   g. Bevel rips can be made in a similar manner except the saw yoke is turned and locked in place as in ripping. Follow the same procedures as discussed in straight ripping.

   h. A bevel-miter (compound angle cut) is a combination bevel and miter cut.

GENERAL SAFETY PRACTICES

1. Wear approved eye protection, hearing protection, and proper clothing. Tie up loose hair and remove loose jewelry.

2. Do not operate the machine without the instructor's permission, or without instructor supervision.

3. Be sure the blade is sharp, sound, and of the proper type.

4. All adjustments should be tight and all guards in place.

5. Never leave tools, scraps, or other materials on the saw table. Keep the area around the saw clear.

6. Do not leave the machine while the blade is rotating.

7. Be sure material is free of knots, nails, or other foreign matter.

8. Do not adjust machine while it is running.

9. Tighten rip lock before starting the saw.

10. Pull saw slowly through material. Return saw to rear of table after sawing. Turn off the power and wait for the blade to stop before removing material.

11. Do not stop blade by pushing material against the blade.

12. Do not saw material freehand without a guide.

COMPLETION QUESTIONS

1. ________________________ or ________________________ types of blades may be used for crosscutting.
2. A ________________________ blade or feeding the saw to ________________________ may result in the saw grabbing the material.
3. The depth of cut into the table is adjusted by turning the ________________________ ________________________.
4. Material should be fed ________________________ the rotation of the blade when ripping.
5. The saw blade should be placed on the arbor so the teeth on the ________________________ of the blade point back toward the elevating column.
6. Be sure all ________________________ are tight before the saw is turned on.
7. Return the saw to the ________________________ of the table after making a crosscut.
8. A ________________________ angle cut is a combination bevel and miter cut.
9. Hold material with your ________________________ hand while crosscutting.
10. To cut a miter, the ________________________ must be released.

SCROLL SAW

PART IDENTIFICATION

Identify the numbered parts of the scroll saw illustrated below.

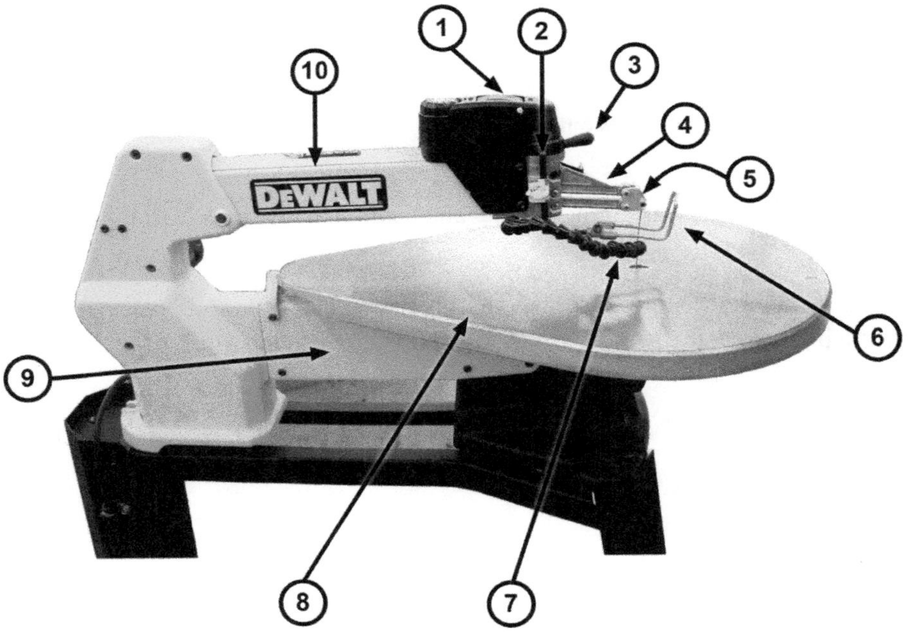

1. ____________________
2. ____________________
3. ____________________
4. ____________________
5. ____________________
6. ____________________
7. ____________________
8. ____________________
9. ____________________
10. ____________________

SAFE OPERATIONAL PROCEDURES

1. Select the right blade for the task. Always have at least 2 teeth in contact with the material. Refer to operator's manual for specific installation of blade type and correct size.
2. Blades vary from 7 teeth per inch for rough cutting to 32 teeth per inch recommended for cutting metal or other hard materials. Blades having 15 teeth per inch are recommended for general purpose cutting.
3. Insert blade with the teeth pointing down so that cutting is on the downward stroke, thus pushing the work against the table.
4. After installing the correct blade adjust the blade tension according to the size of the blade.
5. Adjust saw speed to according to the material being cut; determine proper speed according to the operator's manual.
6. Be sure the guard is in place before operating the saw.
7. Blade alignment is very important for safe and efficient operation. When viewed from the side, the blade should move straight up and down when the saw is running.
8. Adjust the hold-down foot so that the spring tension holds the work tight to the table.

9. Before beginning cut, check the work piece for nails, paint, grit, or other foreign material which could dull the blade or possibly break the blade.
10. With the table clear of all debris and scrap, start the machine and feed the material into the blade forward and evenly with a slight downward pressure.
11. Cuts should be on the waste side of the line. The line should be barely visible on work after cutting.
12. The scroll saw can be used for inside cuts; this is also called a pierce cut.
    a. Drill small holes in the waste material to begin and finish your cut.
    b. Insert the blade through the drilled hole and secure in place.
    c. Proceed with your cut.
    d. When finished with cut, stop machine, raise upper guide, and remove blade to free the work piece.
13. Do not force the material into blade or attempt to turn too sharply. If the blade should break, turn the saw off and allow the blade to come to a complete stop before removing work or replacing the blade.
14. Bevel cuts can be made if the saw is equipped with a tilt table.

GENERAL SAFETY PRACTICES

1. Wear approved eye protection, hearing protection, and proper clothing. Tie up loose hair and remove loose jewelry.
2. Do not operate the machine without the instructor's permission, or without instructor supervision.
3. Use only sharp blades.
4. Keep work area clear and free of debris.
5. All guards/shields must be in place at all times when operating the saw.
6. Make all adjustments before turning on the power.
7. Adjust the hold-down foot so it applies downward pressure to the material.
8. Keep hands and fingers out of line of the saw blade.
9. Do not force material into the saw or attempt to turn too tight of a radius.
10. Do not talk to anyone while using the saw. The operator should be the only person inside the safety zone.

COMPLETION QUESTIONS

1. The saw speed should be adjusted to the material being cut and according to ______________________________ ______________________________.

2. All adjustments should be made with the power__________________________.

3. The work is held snug to the table by the __________________________.

4. Keep hands and __________________________ out of the line of the sawblade.

5. Blades vary from __________________________ teeth per inch for rough cutting to __________________________ teeth per inch recommended for cutting metal or other hard materials.

6. The blade should be inserted so that sawing is completed on the ____________________ stroke.

7. You should __________________________ the strokes per minute based on the material you are cutting.

8. To make a bevel cut you have to __________________________ the table.

9. Saw cuts should be made on the __________________________ side of the line.

10. Blades having __________________________ teeth per inch are recommended for general purpose cutting.

SHAPER

PART IDENTIFICATION

Identify the numbered parts of the shaper illustrated below.

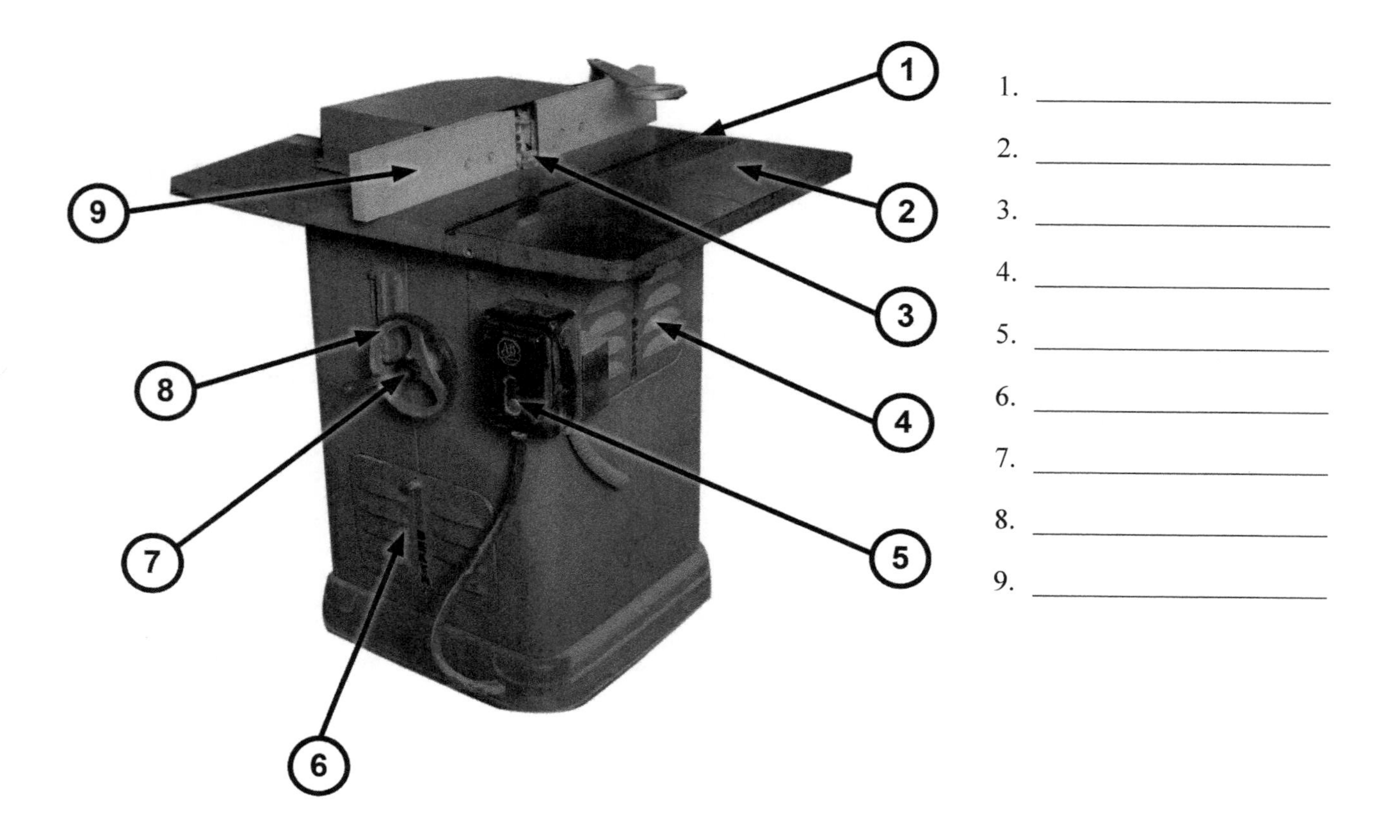

SAFE OPERATIONAL PROCEDURES

1. Study the operation, maintenance, and safety manual for the shaper to be operated.
2. Changing the shaper cutters:
   a. Disconnect the shaper from the energy source.
   b. Select the proper cutter for the job; be sure it is sharp.
   c. Have the proper tools available for loosening and tightening the spindle nut.
   d. Lock the spindle nut according to the manufacturer's procedure.
   e. Insert the cutter so the sharp edge will rotate into the material.
   f. Best results are usually obtained by placing flat material upside down on the table with the cutter on the underside. This procedure makes the operation safer by covering the cutter with the material.

3. Cutting a rabbet, dado, or molding:
   a. Select the correct cutter for the job. Place the proper collars above and below the cutter. For most jobs, the collar is used to control the depth of the cut.
   b. Secure the cutter to the spindle by tightening the nut.
   c. Adjust the height of the spindle and lock in place.
   d. Adjust the fence opening to within 1/8" of the cutter. The fence partially shields the cutter and spindle from the operator.
   e. The fence can be used to control the depth of the cut on straight material. Be sure the fence clears the cutters. At least half of the material must be against the fence at all times.
   f. Turn on motor and make a practice cut on a piece of scrap material.
   g. When making a cut, move the material against the rotation of the cutter.
   h. Cut end grain first to avoid chipping finished corners.
   i. Keep the material moving slowly but steadily to prevent it from burning.
   j. Avoid excess pressure against the collar when it is used for controlling depth of cut to prevent burning. Be sure at least 1/8" of material is against the collar.
   k. Keep hands away from the cutters. Never allow hands to get closer than 12" from the cutter. Use push blocks when working with small pieces.
   l. Turn off the motor after the cut is completed and wait for the motor and cutter to stop before leaving work area.
   m. Remove the cutter and collar from the shaper after the job is completed, and return them to their proper storage places.

GENERAL SAFETY PRACTICES

1. Wear approved eye protection, hearing protection, and proper clothing. Tie up loose hair and remove loose jewelry.
2. Do not operate the machine without the instructor's permission, or without instructor supervision.
3. Use only sharp cutters and be sure they are mounted properly.
4. Be sure the shaper is disconnected from the energy source when changing cutters or making any adjustments.
5. Never start the shaper when the cutter is in contact with the material to be cut.
6. Do not talk to anyone while operating the shaper.
7. Keep the work area free of scrap material; do not try to remove small pieces from the area near the cutter until all mechanical motion stops.

COMPLETION QUESTIONS

1. The cutter is held in place by a ____________________________ threaded onto a short shaft called a ____________________________.
2. The cutter must be mounted so its sharp edge will rotate ____________________________ the material.
3. The material must be fed into the cutter ________________________ the rotation.
4. The ____________________________ is used to control the depth of cut if the edges of the material are not straight.
5. The ____________________________ can be used to control the depth of cut on straight sided material.
6. ____________________________ grain should be cut first to avoid chipping the finished corners.
7. If the movement of the material is too slow or stopped while in contact with the rotating cutter or collar, ____________________________ of the material will result.
8. When shaping flat material, best results will usually be obtained if the collar is mounted ____________________________ the cutter.
9. When making ____________________________ follow the operators' manual instructions for locking the spindle.
10. Never allow hands to get closer than ____________________________ inches from the cutters.

SPOT/RESISTANCE WELDER

PART IDENTIFICATION

Identify the circled parts on the spot/resistance welder illustrated below.

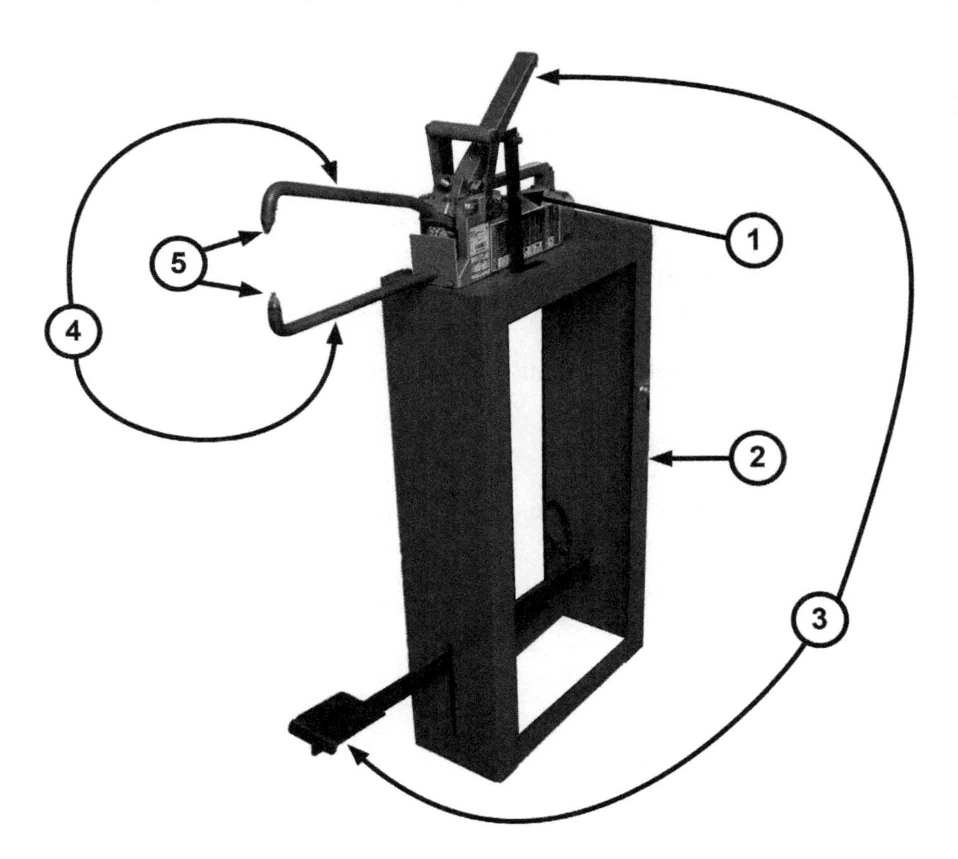

1. ____________________
2. ____________________
3. ____________________
4. ____________________
5. ____________________

SAFE OPERATIONAL PROCEDURES

1. Make sure the material to be spot welded is clean and dry.
2. Make sure the workspace is ready for the work to be done.
3. Do not touch the tongs or tips of the spot welder while getting your material into place.
4. Do not run the machine without metal between the tips/tongs.
5. Avoid touching the spot weld until it has cooled completely.
6. Use a pliers to handle metal that has been welded on.

GENERAL SAFETY PRACTICES

1. Wear approved eye protection, hearing protection, and proper clothing. Tie up loose hair and remove loose jewelry.

2. Do not operate the machine without the instructor's permission, or without instructor supervision.
3. Work in a clean, well-lit environment with proper ventilation.
4. Ensure the workspace is free of debris and moisture, do not operate the spot welder in the presence of water.
5. Do not touch the tips of the spot welder or the material that has been spot welded; they will be hot and burn on contact.
6. Do not leave the machine unattended, disconnect the power source when the work has been completed and when it is not in use.
7. Metal to be joined by spot welding should be free of dirt, oil, water, and other chemicals.
8. Do not spot weld on galvanized metal to prevent toxic fumes, if necessary to spot weld on galvanized material remove the galvanized zinc coating first.
9. Do not allow current to pass through the spot welder without metal between the tips/tongs.

COMPLETION QUESTIONS

1. _______________ _______________ should be worn when spot welding to prevent sparks from flying into the face or eyes.
2. Never run the machine/allow current to pass through the tips/tongs without _______________ between the tips/tongs.
3. Make sure the tips are properly _______________.
4. Work should only be done in a clean, well-lit and properly _______________ area.
5. Do not _______________ the machine without the instructor's permission, or without instructor supervision.
6. Avoid touching the spot weld until it has _______________ completely.
7. Use a pliers to handle _______________ that has been welded on.
8. Metal needs to be clean and free of dirt, _______________, water and other chemicals.
9. Welding on galvanized material can release _______________ fumes.
10. When the work is finished _______________ the power source of the spot welder before leaving it.

TABLE SAW

PART IDENTIFICATION

Identify the numbered parts of the table saw illustrated below.

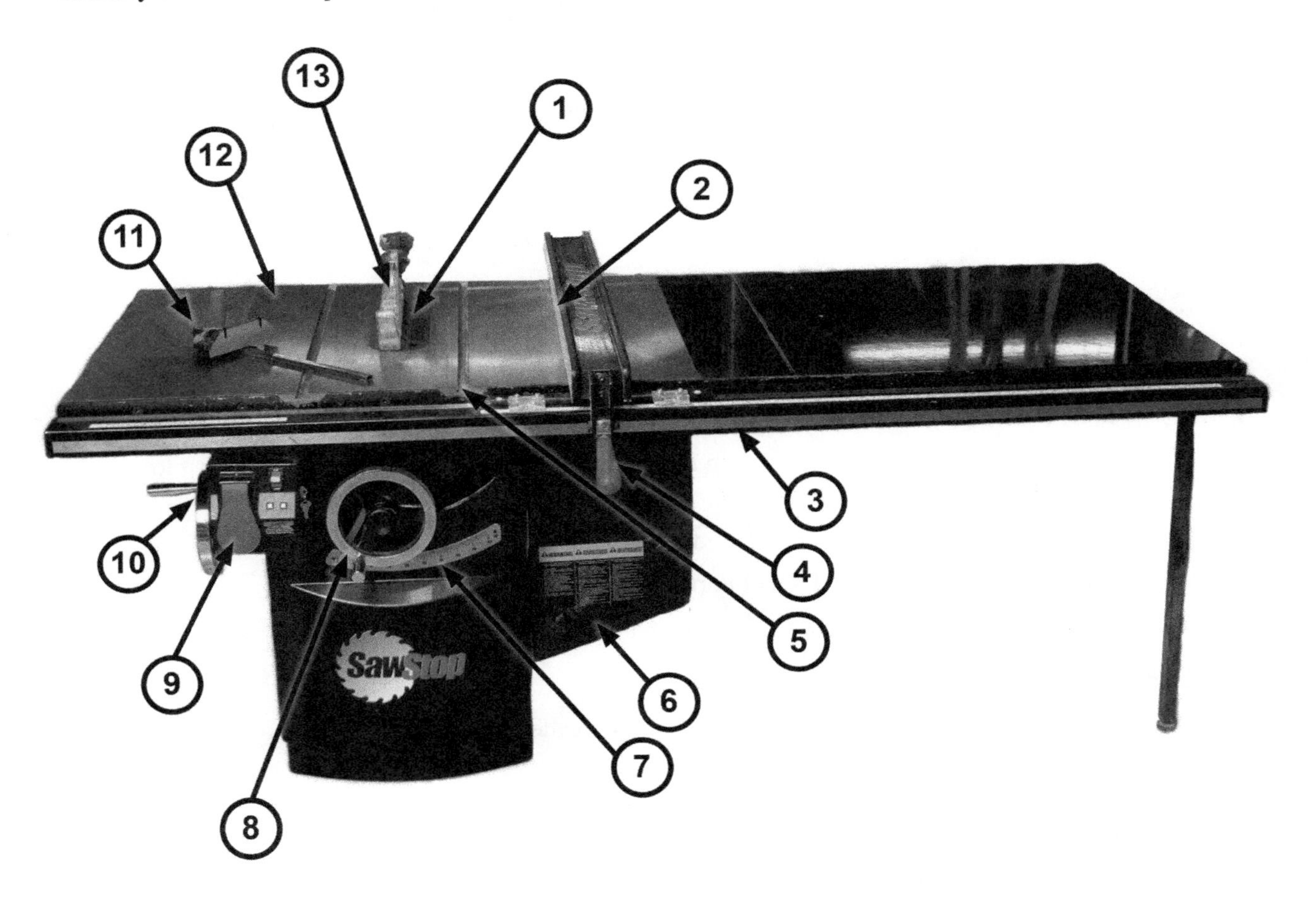

1. ______________________________
2. ______________________________
3. ______________________________
4. ______________________________
5. ______________________________
6. ______________________________
7. ______________________________
8. ______________________________
9. ______________________________
10. ______________________________
11. ______________________________
12. ______________________________
13. ______________________________

SAFE OPERATIONAL PROCEDURES

1. Crosscutting:
   a. Use a crosscut or combination blade.
   b. Adjust the height of the blade so that the teeth clear the thickness of material by ¼”.
   c. Be sure the blade guard and riving knife/splitter are in place. Check the anti-kickback device to make sure it only allows the material to move from the infeed side to the outfeed side of the saw.
   d. Always use the miter gauge when crosscutting. The edge of the material that sets against the miter gauge needs to be straight, true, and smooth.
   e. To prevent kickbacks, never use the ripping fence as a guide when crosscutting.
   f. Place material against miter gauge and cut the material, saving the cutting line.
2. Ripping:
   a. Use a ripping or combination blade.
   b. Keep your material tight to the rip fence while cutting. The edge of the material that sets against the rip fence needs to be straight, true, and smooth.
   c. Check the cutting width by measuring the distance between side of the tooth closest to the fence and the fence.
   d. When ripping material less than 4” wide, use a push stick.
   e. Use a helper or a roller stand to support long pieces of material while sawing.
   f. Adjust the height of the blade so that the teeth clear the thickness of material by ¼”.
3. Dados and rabbets:
   a. Use a dado blade stack/set if available. A single blade may be used instead by making multiple passes next to each other.
   b. Raise the blade to the desired depth of cut. Test your cut on scrap of the same material before cutting into the final work piece.
   c. Making a rabbet cut on the edge of a board may require the removal of the blade guard. Using a feather board will make this operation safer.
4. Bevel cuts (ripping):
   a. Use a combination or ripping blade.
   b. Adjust saw by tilting the blade to the desired angle.
   c. Adjust the height of the blade so that the teeth clear the thickness of material by ¼”.
   d. Adjust ripping fence to desired width of cut.
5. Bevel cuts (crosscutting):
   a. Use a combination or crosscutting blade.
   b. Use the miter gauge, not the ripping fence.
   c. Adjust saw by tilting the arbor and the blade to the angle desired.
   d. Adjust the height of the blade so that the teeth clear the thickness of material by ¼”.

GENERAL SAFETY PRACTICES

1. Wear approved eye protection, hearing protection, and proper clothing. Tie up loose hair and remove loose jewelry.
2. Do not operate the machine without the instructor's permission, or without instructor supervision.
3. Use only sharp blades of the proper type for the job. Make sure the blade is always cutting down toward the front or operator's side of table.
4. Do not stand in line with the blade while operating or allow fingers or hands to be in the line of the blade.
5. Be sure that all adjustments are tight and the table is free of tools, debris, or other materials. Be sure guards are in place for all sawing operations. Make sure the blade has stopped before adjusting the guard.
6. Do not force material into the blade. If the blade binds or overheats, turn the power off immediately.
7. Do not talk to anyone while using the saw. The operator should be the only person inside the safety zone.
8. Be sure the floor is clean and free from scraps and debris. Do not work on wet or slippery floors. Non-skid materials are recommended.
9. Saw only material that has a straight and true edge and is flat on the face touching the table.
10. Understand how to make all adjustments on the saw prior to operation.
11. Hold material against the ripping fence when ripping and the miter gauge when crosscutting. Never make freehand cuts.
12. Be sure material clears the blade before reaching to turn the power off; do not move while the blade coasts to a stop. Only remove scrap once the blade has stopped.
13. Do not place the hands over or in front of the blade. Never reach over the blade.
14. The power should be off before making any adjustments. More difficult adjustments such as changing the blade should only be made with the power source removed from the saw.

COMPLETION QUESTIONS

1. A ______________________________ or ______________________________ is used to support long pieces of material while sawing.
2. The ______________________________ is used for a guide when ripping.
3. The ______________________________ is used as a guide when crosscutting.
4. A ______________________________ or ______________________________ blade may be used for crosscutting.

5. A ____________________________ is used when ripping pieces narrower than 4”.

6. When making a bevel cut, the saw is adjusted by tilting the ____________________________.

7. Saw only material that has a ____________________________, smooth, and true edge.

8. The blade should be ____________________________ before removing scrap from the blade.

9. A ____________________________ or a single blade can be used when cutting dados.

10. The ____________________________ should be raised above the material being cut by ¼”.

TIRE BALANCING MACHINE

PART IDENTIFICATION

Identify the circled parts on the tire balancing machine illustrated below.

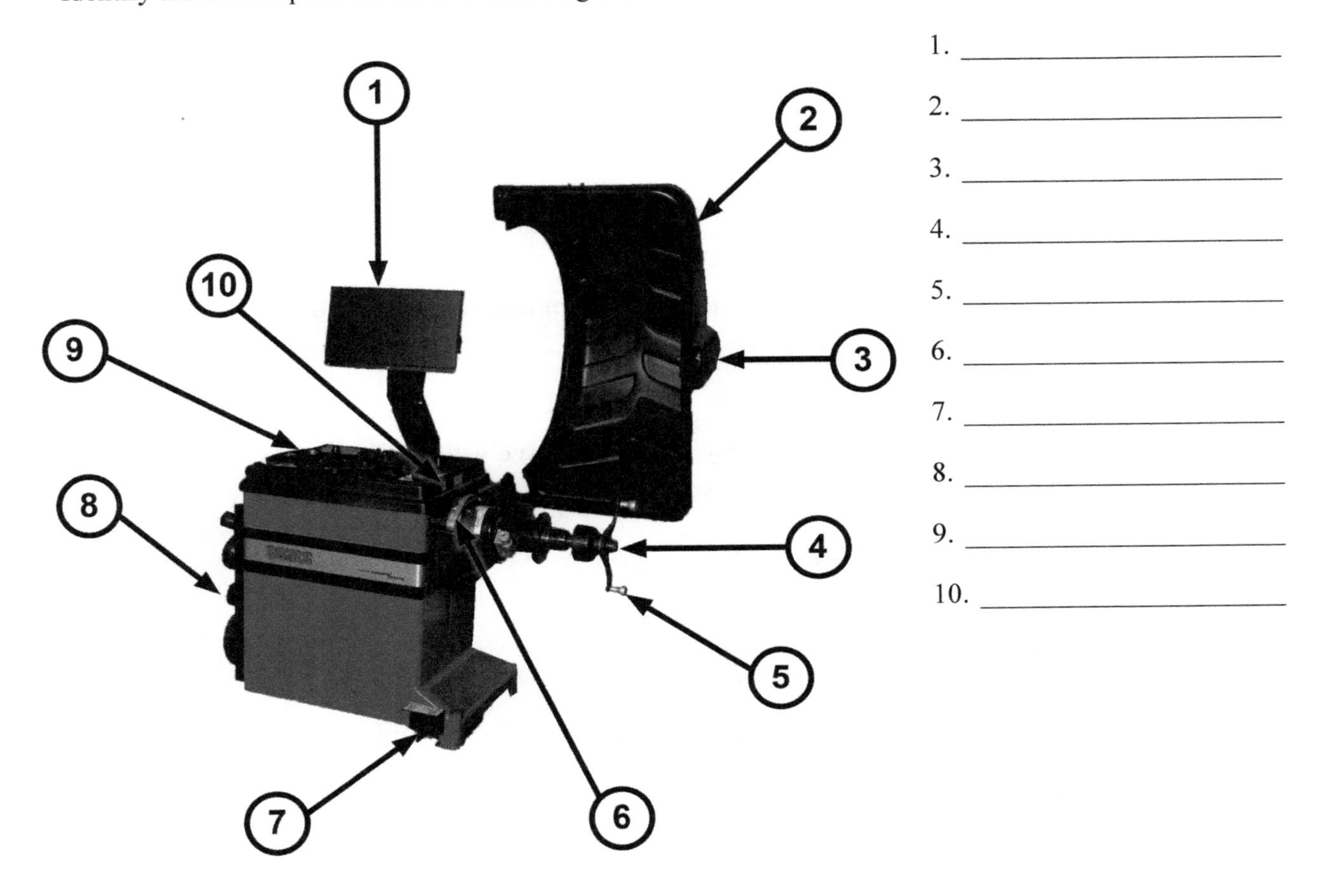

SAFE OPERATIONAL PROCEDURES

1. Make sure the machine is securely bolted to the floor.
2. Tire balancing machine will need to be calibrated after initial installation.
3. Disconnect the tire balancing machine from the power source when not in use.
4. Inspect and maintain your machine on a regular basis, look for moisture, dirt and rust.
5. Check to be sure all tools and accessories are clean and in good working order.
6. Remove any foreign material from the tire and rim before mounting such as dirt, mud, and snow because it will affect the balance of the tire.
7. Inspect the tire and rim carefully for wear and defects before spinning.

8. Position your body to the side and keep hands & fingers away from the rotating tire area.
9. Be cautious when the tire is rotating on the tire balance machine.
10. Ensure the rim and tire size do not exceed the machine's capacity.
11. Never remove the tire hood from the machine.
12. Make sure the tire hood is down when spin balancing a tire.
13. Use the wheel weight hammer/plier to detach and apply any wheel weights on the rim.

GENERAL SAFETY PRACTICES

1. Wear approved eye protection, hearing protection, and proper clothing. Tie up loose hair and remove loose jewelry.
2. Do not operate the machine without the instructor's permission, or without instructor supervision.
3. Safety toe boots are highly recommended.
4. Maintain a constant awareness of the many hazards involved with balancing tires.
5. Never allow unqualified persons in the area while you are working.
6. Check the rotating spindle and collets for corrosion.
7. Ensure the locking devices (clamps) are properly working.
8. Never inflate a tire past its recommended air pressure.
9. Never try to repair the tire machine, report any problems to your instructor.

COMPLETION QUESTIONS

1. Stand to the ________________ when a tire balancing machine is spinning.
2. ________________, ________________, or ________________ in the rim will affect the end result when balancing a tire.
3. It's important to check the tire's air ________________ before balancing it.
4. Aluminum and steel rims ________________ take the same kind of wheel weights.
5. A tire balancing machine must be ________________ to the floor.
6. A tire balancing machine may have to be ___________ after its original installation.
7. Always wear your ________________ ________________ ________________ when operating the balance machine.
8. Never remove the ________________ ________________ on the tire balancing machine.
9. Use the correct wheel ________________ and apply them correctly.
10. Be sure the tire and rim are the ________________ size for the machine.

TIRE CHANGER/MOUNTING MACHINE

PART IDENTIFICATION

Identify the circled parts on the tire changer/mounting machine illustrated below.

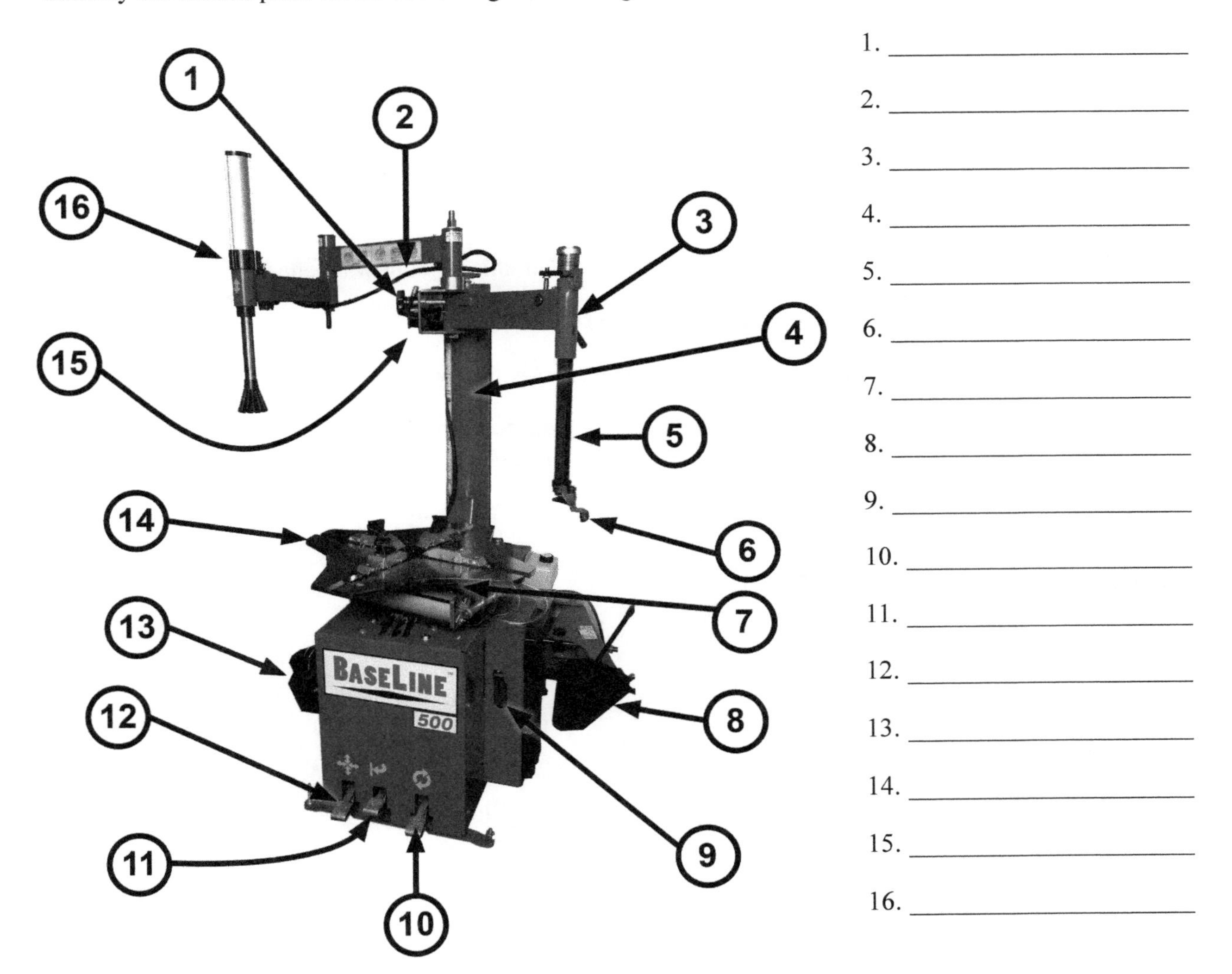

SAFE OPERATIONAL PROCEDURES

1. Follow all safety procedures as defined by the manufacturer in the safety manual.
2. Inspect your machines on a regular basis, look for air leaks, oil leaks, and make sure it is securely bolted to the floor.
3. Disconnect the power source from the wall before servicing the machine.
4. Check to be sure all tools and accessories are clean and in good working order.

5. Position your body to be in the middle of the table and keep hands & fingers away from the bead seat area.
6. Be cautious when the tire is rotating on the tire machine table.
7. Inspect tire and rim carefully for wear and defects before mounting.
8. Be sure the tire and rim are the same size and in good condition.
9. Deflate the air in the tire completely by pulling out the valve core.
10. Use the wheel weight hammer/plier to detach any wheel weights on the rim.
11. Don't force the issue, breaking the bead can be a problem, use the proper tools.
12. You may need to lubricate the bead to get the tire off the rim.
13. When mounting a tire, use the tire lubricant as needed, this will allow a better seal.
14. Rotate the turning table to spin the tire onto the rim.
15. Install the valve core, release the tire from the turning table.
16. Inflate the tire to the manufacturer's recommended air pressure. Inflate the tire with short bursts of air, checking pressure regularly, and do not stand directly over the tire.
17. Make sure all tools and accessories are stored away, and all equipment is cleaned up.

GENERAL SAFETY PRACTICES

1. Wear approved eye protection, hearing protection, and proper clothing. Tie up loose hair and remove loose jewelry.
2. Do not operate the machine without the instructor's permission, or without instructor supervision.
3. Work in a well-ventilated area when using a patching compound.
4. Safety toe boots are highly recommended.
5. Maintain a constant awareness of the many hazards involved with changing tires.
6. There are many pinch points that are capable of inflicting injury from the tire mounting machine.
7. Never allow unqualified persons in the area while you are working.
8. When repairing a rim by welding, deflate and dismount the tire.
9. Check the table clamps, rotating arm for corrosion, damage or sticky operation.
10. Ensure the locking devices (clamps) are properly working.
11. Never try to repair the tire machine, report any problems to your instructor

COMPLETION QUESTIONS

1. Do not operate the machine without the _______________ permission.
2. Inspect your machines on a _______________ basis.
3. It is necessary to use some sort of _______________ on the tire when removing/installing tires.
4. Many _______________ _______________ are possible when using a tire machine.
5. Manufacturers recommend the maximum _______________ _______________ for the tire you are working on.
6. Make sure you follow the _______________ safety procedures when using the tire machine.
7. Never mount or balance a _______________ tire and wheel.
8. Do not stand _______________ the tire while inflating with air.
9. The tire mounting machine must be _______________ to the floor.
10. Make sure all _______________ and _______________ are stored away, and all equipment is cleaned up.

VERTICAL BAND SAW

PART IDENTIFICATION

Identify the numbered parts of the vertical band saw illustrated below.

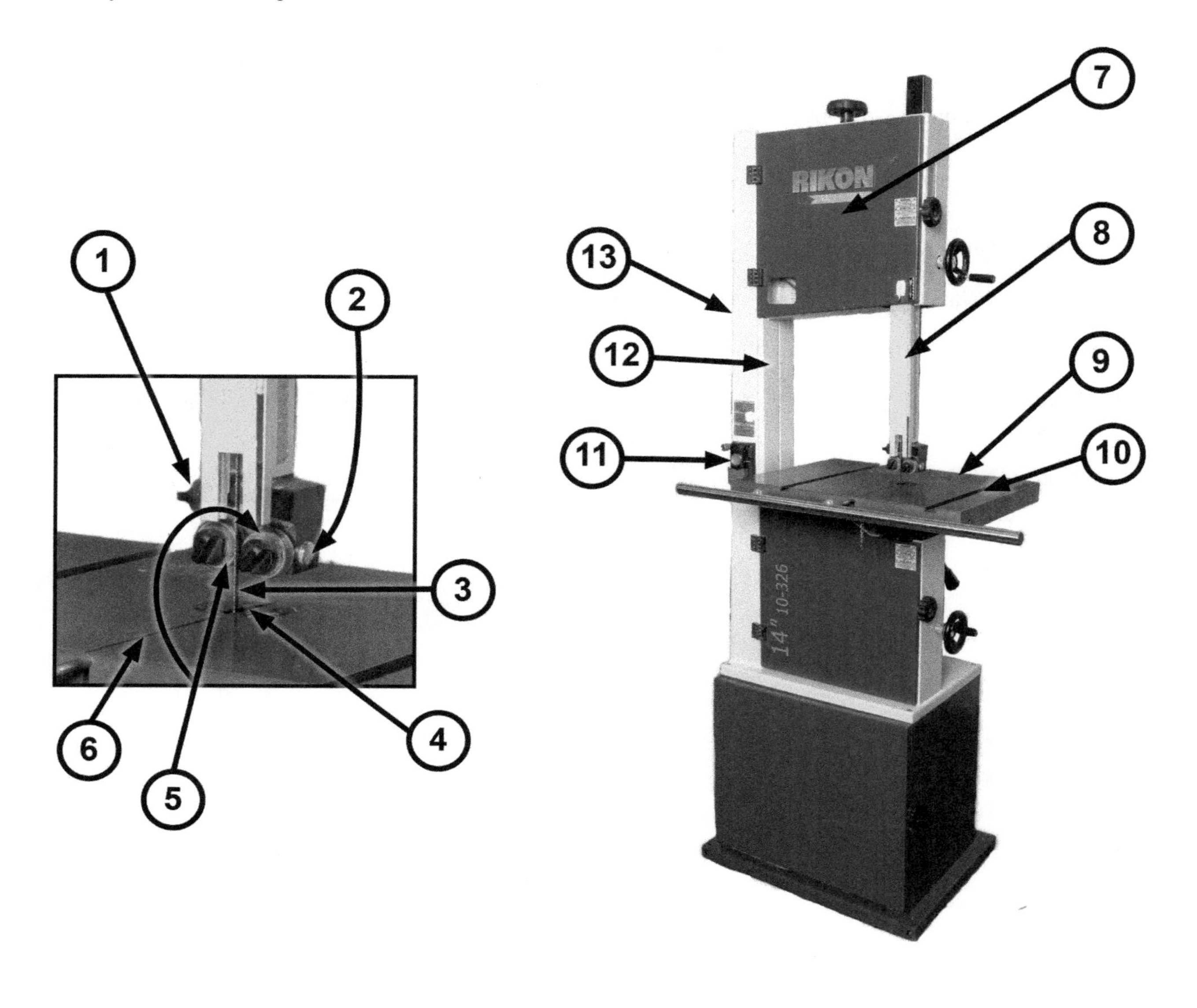

1. ______________________________

2. ______________________________

3. ______________________________

4. ______________________________

5. ______________________________

6. ______________________________

7. ______________________________

8. ______________________________

9. ____________________

10. ____________________

11. ____________________

12. ____________________

13. ____________________

SAFE OPERATIONAL PROCEDURES

1. Adjust the upper guard and guide 1/8" to 1/4" above the material to be cut.
2. Select proper blade width. General rules regarding minimum radius cuts are determined by the operator's manual.
3. Only use clean, sharp blades.
4. Replace blades upon signs of wear.
5. Blade guides or rollers should be set a paper thickness away from the blade.
6. Teeth on the blade should point down in the direction of travel of the blade.
7. Check tension of the blade frequently with the power off and saw unplugged. Make adjustments as necessary; check the tracking of the blade by turning the upper wheel by hand.
8. Adjust the vertical band saw blade so that it will run on the center of the wheels and straight through the guides.
9. When not cutting, the blade should run 1/32 of an inch ahead of the thrust/blade support wheel.
10. Be sure the material is free of nails, paint, and defects.
11. Make sure the material is not in contact with the blade when starting the machine.
12. Avoid backing out long cuts or curves when possible while the machine is running.
13. Make relief cuts before cutting curves.
14. Do not start cutting until the saw is running at full speed.
15. Feed the material into the saw slowly. Do not force material.
16. If freehand sawing, use one hand to guide the work and the other to push the work into the saw.
17. Do not twist/turn the material too sharply and produce excess stress on the blade which could cause the blade to break.
18. If the material binds or pinches the blade, do not attempt to back out until the power has been shut off and the machine stops.
19. Use the fence or miter gauge when cutting straight lines.

20. The table can be tilted for making bevel cuts.

21. Cylindrical material should be mounted in a V-block to keep it from spinning.

GENERAL SAFETY PRACTICES:

1. Wear approved eye protection, hearing protection, and proper clothing. Tie up loose hair and remove loose jewelry.
2. Do not operate the machine without the instructor's permission, or without instructor supervision.
3. Keep floor and surrounding area free of debris that might cause tripping.
4. Keep all guards in place at all times.
5. Have someone assist in operations which are not safely handled alone.
6. Make all adjustments with the power off and blade stopped.
7. Keep hands a safe distance from moving parts, never closer than 2" from the blade.
8. Give undivided attention to the job. The operator should be the only one inside the safety zone area.
9. Use a push block when cutting small parts.
10. Never reach around a moving blade.
11. When making a cut, do not place hands in line with the cutting line or blade.
12. Never attempt to remove scrap or debris while the power is on or the blade is running.
13. When finished cutting, turn off the power. Do not leave the machine until the blade comes to a complete stop.

COMPLETION QUESTIONS

1. Only use ________________________ blades that have been adjusted properly.
2. When the blade is properly installed, the teeth should point ________________________.
3. The blade should be adjusted so it runs ________________________ of an inch from the thrust/blade support wheel.
4. Keep hands at least ________________________ inches from the blade.
5. Use a ________________________ when cutting small parts.
6. Adjust the upper guide about ________________________ above the material being cut.
7. Properly adjust the vertical band saw blade so it will run on the ________________________ of the wheels.

8. Do not make any adjustments unless the power is off and the blade has come to a complete ________________________.
9. Never back out of a cut with the blade ________________________.
10. Never ________________________ the blade too sharply because it could cause the blade to break.

WIDE BELT SANDER (TIMESAVER)

PART IDENTIFICATION

Identify the circled parts on the wide belt sander (timesaver) illustrated below.

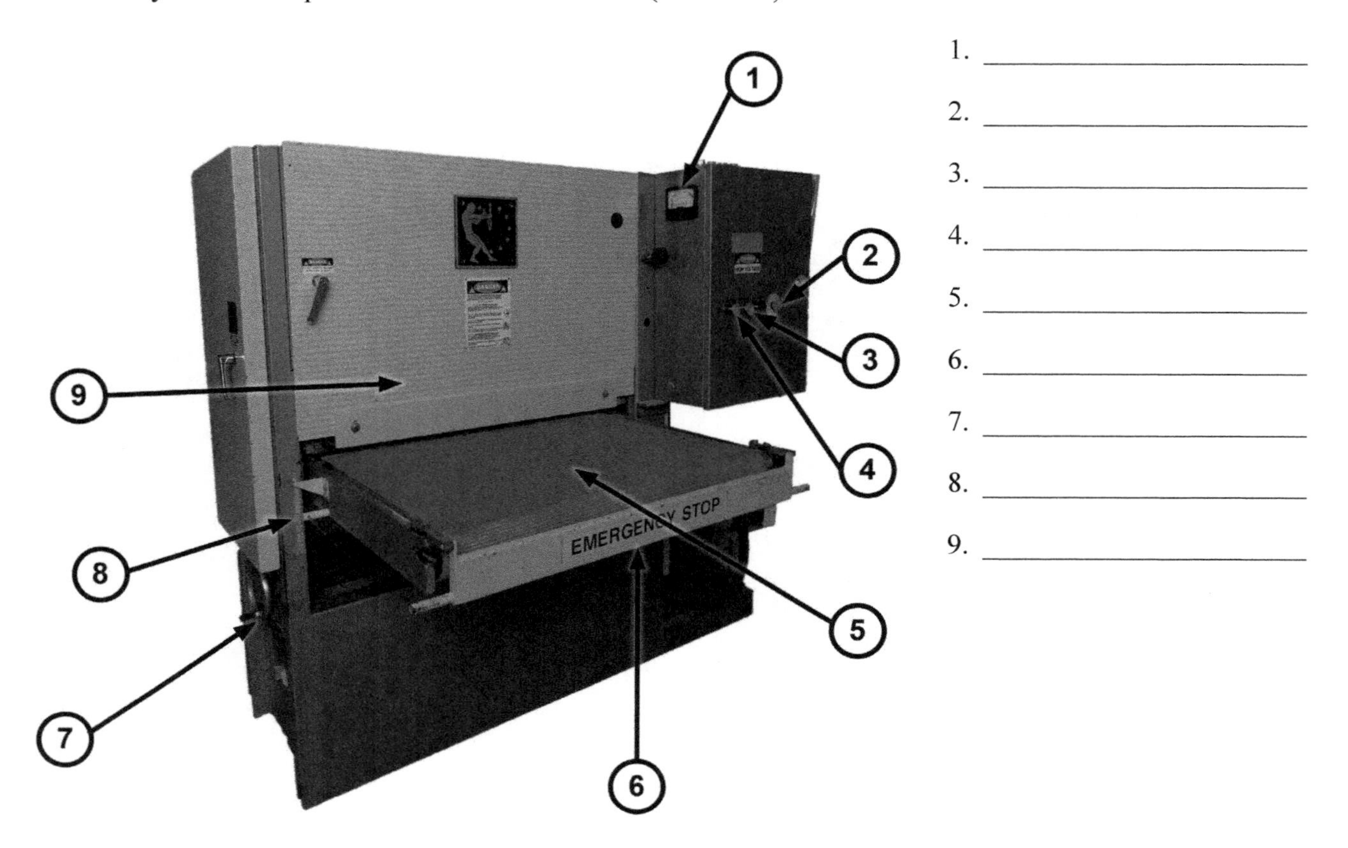

1. ______________________
2. ______________________
3. ______________________
4. ______________________
5. ______________________
6. ______________________
7. ______________________
8. ______________________
9. ______________________

SAFE OPERATIONAL PROCEDURES

1. Ensure your material is free of any metal fasteners, and remove as much excess glue as possible with a scraper. The glue will cause the sandpaper to “load up” and burn your workpiece.
2. Turn the dust collector on and open any blast gates.
3. Ensure the compressed air is on and maintaining pressure.
4. Clear anyone from standing directly behind the machine.
5. Clear the table of all materials and tools. The table is a conveyer belt that will feed your material into the sander.
6. Locate the emergency stops in case you need to shut the sander down suddenly.

7. Measure your material to ensure it is longer than 12" when measuring the long side of the grain. If it is shorter, it could get stuck between the drive rollers, ruining your workpiece.

8. Measure the thickness of the material to be sanded in multiple places with a caliper. It must be thicker than ⅛".

9. Set the table to the same measurement as the thickest part of your material.

10. Turn the conveyor on and ensure it is staying centered on the rollers.

11. Adjust the feed rate of the conveyor. Remember the faster you adjust the conveyor, the less material you can take off at a time.

12. Standing to the side, turn the sanding head on and allow it to come to full speed.

13. Keeping your hands from under the material, place the flattest side of your material on the conveyor belt allowing the machine to feed the material. Do not push the material through the sander.

14. Remove the material from the back of the sander and inspect for quality of finish.

15. If there are spots not sanded or you are trying to achieve the desired thickness, raise the table according to the material and abrasive you are using and repeat passes. Mark with a pencil all over your board. When all the pencil marks are gone, the desired amount has been taken off and your board should be smooth.

16. Shut the machine down and wait for all motion to stop before leaving.

GENERAL SAFETY PRACTICES

1. Wear approved eye protection, hearing protection, and proper clothing. Tie up loose hair and remove loose jewelry.

2. Do not operate the machine without the instructor's permission, or without instructor supervision.

3. Sanding produces fine dust and respiratory protection should be worn to prevent inhalation of wood dust.

4. This sander is designed and intended for use by properly trained and experienced personnel only. If you are not familiar with the proper and safe operation of this sander, do not use it until proper training and knowledge has been obtained.

5. Keep safety guards in place at all times when the machine is in use. If removed for maintenance purposes, use extreme caution and replace the guards immediately.

6. Give your work undivided attention. Looking around, carrying on a conversation, and "horse-play" are careless acts that can result in serious injury.

7. Remove loose items and unnecessary material from the area before starting the machine.

8. Keep hands clear while feeding workpieces onto the conveyor table. The workpiece will be forced down as it begins to feed into the machine, causing a pinching action between the material and conveyor table so be sure fingers don't get pinched between your work and the table.

9. Stand to one side of the conveyor table and do not let anyone else stand in line with the table while a workpiece is being fed through the machine.
10. Never reach into a running machine. Turn off and disconnect from the energy source before attempting to retrieve parts or material from within the machine.
11. Never leave the machine running unattended. Turn the power off and do not leave the machine until all parts come to a complete stop.
12. If you are sanding long materials, provide support with a helper or a roller stand.
13. The sander will feed material into the machine itself. If the sander slows down, you are removing too much material. Keep a careful watch on the main motor load meter. If it jumps into the red zone, turn the machine off immediately.

COMPLETION QUESTIONS

1. Ensure the _______________ _______________ is on and all blast gates are open.
2. Clear anyone from standing directly _______________ the machine.
3. Locate the _______________ _______________ in case you need to shut the sander down suddenly.
4. Set the table to the _______________ measurement as the thickest part of your material.
5. Measure your material to ensure it is longer than _______________ when measuring the long side of the grain.
6. Remember the _______________ you adjust the conveyor, the _______________ material you can take off at a time.
7. Sanding produces _______________ _______________ and respiratory protection should be worn.
8. Do not attempt to sand stock shorter than _______________ long. Do not sand stock less than _______________ thick.
9. The workpiece will be forced _______________ as it begins to feed into the machine, causing a pinching action between the material and conveyor table.
10. The sander will feed material into the machine itself. If the sander slows down or the main motor load meter jumps into the red, you are removing _______________ _______________ material.

WOOD LATHE

PART IDENTIFICATION

Identify the numbered parts of the wood lathe illustrated below.

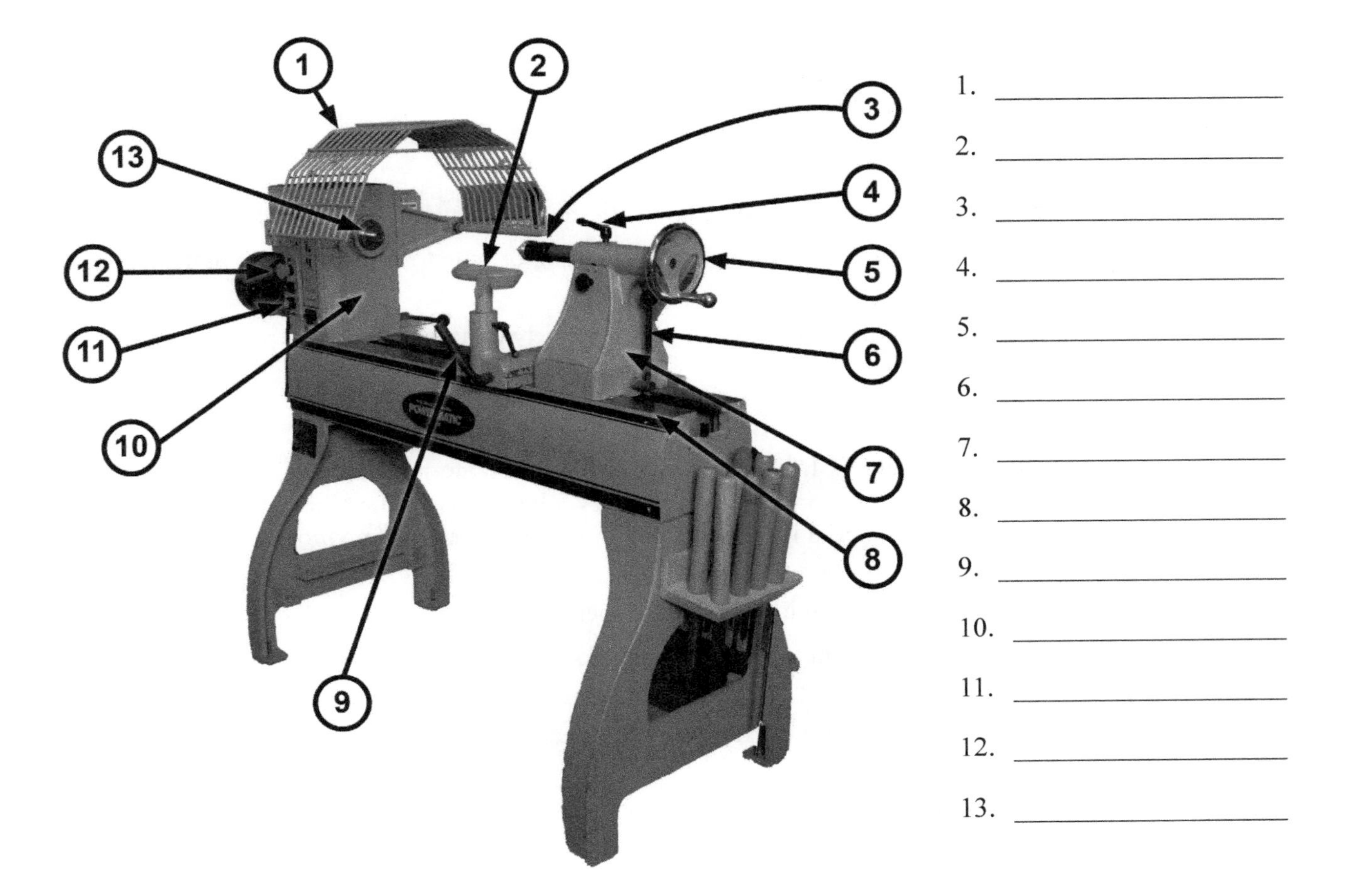

SAFE OPERATIONAL PROCEDURES

1. The wood lathe should be considered one of the more dangerous items in the laboratory/shop due to the fast-moving material that cannot be fully guarded.
2. Different speeds are required for different lathe operations. Refer to the operator's manual to determine the safe speeds.
3. Spindle turning or between-center turning:
   a. In spindle turning or between-center turning, the material is mounted between centers of the head-stock and tail-stock.
   b. Select a piece of material approximately one inch longer than the finished length. It should be free of all wood defects.
   c. If larger than 2" square, saw or plane off the corners.

d. Square the ends and find the center of each end by drawing diagonal intersecting lines.

e. Select one end as the live center and cut saw kerfs about 1/8" deep along the diagonal lines. Drive the spur (headstock) center into the kerfs with a mallet.

f. On the other end, at the point where the lines intersect, center punch a hole approximately 1/8" in diameter and 3/16" to 1/4" deep.

g. Replace the live center spur into the headstock.

h. Adjust the tailstock spindle

(1) Advance tailstock 1" beyond the tailstock housing.

(2) Slide the tailstock until the point of the dead center enters the hole in the material.

(3) Lock the tailstock in place.

(4) Turn the tailstock spindle feed handle until the dead center is seated in the material.

(5) Release the pressure slightly and apply oil, wax, or soap to the impression in the end of the material.

(6) Turn the feed handle until the dead center is fairly tight in the original position.

(7) Back off slightly until the material rotates freely on the dead center.

(8) Lock the spindle in place.

i. Adjust the tool rest until it is 1/8" above the center of the material and there is a 1/8" clearance between the rest and the widest portion of the material. The tool rest should be moved forward as the material is reduced in size with a clearance of no more than 3/8" at any time.

j. Select a gouge tool for the rough work. Remember cutting tools must be sharp.

k. For the rough work, the machine should run at a low speed according to operator's manual.

l. Assume a natural position with the feet slightly spread with one foot slightly ahead of the other.

m. Grasp the handle of the tool well toward the end with the dominate hand. Hold the blade with the other hand guiding it along the tool rest. Work from the center toward the end.

n. Continue turning until the material is cylindrical and near the finished size. Finish the turning job with a skew cutter. The speed may be increased according to the operator's manual for smoother cutting.

o. Depending on the completed job desired, select other cutters accordingly.

p. When the job is complete with the cutters, remove the tool rest for sanding. Select proper sandpaper and hold a long piece between both hands, moving the paper back and forth along the work. Don't touch the rotating material with your hand.

4. Headstock or faceplate turning:

a. In headstock or faceplate turning, the material is mounted to a flat metal plate which is attached directly to the headstock.

b. Band saw a round disc at least 1/8" larger than finished dimension.

c. Attach a faceplate smaller than the material with sufficient wood screws, ensure clearance from the length of the screws so they will not be cut into.

d. If turning laminated or glued material, make sure the glue joints are secure. Never turn the lathe to its top speed on laminated material.

e. Remove the spur center and screw the faceplate and material in the headstock spindle.

f. Adjust the tool rest parallel to the center of the material and about 1/8" from the face of the material.

g. Select the slowest speed on the lathe. With the power off, turn the material by hand to see that it clears the tool rest.

h. Rough turn the outside of the material with a gouge, and then smooth with a round nose tool.

i. Reset the tool rest so it is parallel to the center surface of the material. Using a round nose cutter, remove the inside or center of the material on the downward side. Continue turning until the bottom and walls are to the desired thickness, never less than 3/8" on the bottom or 1/4" on the wall thickness.

j. For sanding, remove the tool rest and fold a medium or fine sandpaper into a pad for smoothing the material. Use a lathe speed according to operator's manual.

GENERAL SAFETY PRACTICES

1. Wear approved eye protection, hearing protection, and proper clothing. Tie up loose hair and remove loose jewelry.
2. Do not operate the machine without the instructor's permission, or without instructor supervision.
3. Check for clearance: Spin work by hand before turning on power.
4. Always use sharp tools and the proper tool for the job.
5. Make sure tool rest is approximately 1/8" from widest portion of the material and no more than 3/8" away at any time.
6. Remove tool rest before sanding.
7. Do all rough cutting at low speed.
8. Avoid heavy cuts.
9. Hold cutting tool firmly and at proper angle to work.
10. Don't leave tools on bed of lathe.
11. Stop the lathe often to check tailstock adjustment.
12. Keep tailstock (dead center) well lubricated.
13. Make all adjustments with power off and disconnected from the energy source.
14. Select proper and safe speeds for all operations according to operator's manual.
15. Keep work area clean and free of chips and shavings at all times.
16. Never touch material while it is turning no matter how smooth it may seem.

17. Do not talk to anyone while operating the lathe.
18. All loose hair and clothes needs to be tied up.

COMPLETION QUESTIONS

1. All work turned in the lathe should be started with a ______________________ speed.
2. Before starting the lathe, always turn the material by hand to see that it clears the ______________________ ______________________.
3. To turn a small bowl, the material is attached to the ______________________.
4. The center of the material is determined by drawing ____________________ lines and the center is at the ______________________ of the two lines.
5. Most tailstock spindles remain solid and are called ______________________ centers.
6. Tie up loose ______________________.
7. When beginning a job, the tool rest should be set at ____________________ from the material and never allowed to become more than ______________________ from the material as the turning operation progresses.
8. A ______________________ is the cutting tool commonly used for rough cutting.
9. The tailstock should be ________________________ from time to time to keep it from burning or causing the material to lock up in the lathe.
10. The ______________________ or headstock center is driven into the material before inserting into the lathe.

INDEX